AF764960

Plant Diseases: Identification and Management with Illustrations

Plant Diseases: Identification and Management with Illustrations

Editor

Cornelio Losa

scitus academics

Plant Diseases: Identification and Management with Illustrations
Edited by **Cornelio Losa**

ISBN: 978-1-68117-579-9
Library of Congress Control Number: 2016946433

© 2017 by
SCITUS Academics LLC,
616, Corporate Way, Suite 2, 4766,
Valley Cottage, NY 10989

www.scitusacademics.com

This book contains information obtained from highly regarded resources. Copyright for individual articles remains with the authors as indicated. All chapters are distributed under the terms of the Creative Commons Attribution License, which permits unrestricted use, distribution, and reproduction in any medium, provided the original author and source are credited.

Notice

Reasonable efforts have been made to publish reliable data and views articulated in the chapters are those of the individual contributors, and not necessarily those of the editors or publishers. Editors or publishers are not responsible for the accuracy of the information in the published chapters or consequences of their use. The publisher believes no responsibility for any damage or grievance to the persons or property arising out of the use of any materials, instructions, methods or thoughts in the book. The editors and the publisher have attempted to trace the copyright holders of all material reproduced in this publication and apologize to copyright holders if permission has not been obtained. If any copyright holder has not been acknowledged, please write to us so we may rectify.

Printed in United States of America on Acid Free Paper

Preface

All plant species are controlled by disease, both wild and cultivated. While every species is prone to characteristic diseases, they are relatively few in number in every case. Diseases are caused by certain disturbances in the plant's standard life process. Diseases might be the result of life and/or non-living causes. Biotic diseases are caused by living organisms (for example, bacteria, fungi and viruses). Abiotic diseases are caused by non-living environmental conditions (e.g. compaction of soil, wind, frost, damage to soil salt and roots of girdle). Weather plays a major role in the development of fungal diseases. Plant diseases have caused severe human loss in a number of ways. The objective of plant disease management is to reduce the esthetic and economic damage caused by plant diseases. This has traditionally been referred to as plant disease control, but it has social and environmental values that consider control to be absolute and the term too rigid. More versatile approaches to disease management and integrated disease management have resulted from this change in attitude, but individual measures, often intense, such as pesticide applications, soil fumigation or perhaps burning, are not commonly used.

This book covers state of the art real-world practices and knowledgeable information that helps to manage diseases in food, fiber and landscape plants successfully. Information on; the accurate identification of diseases and the pathogens that cause them; the development of plant disease epidemics; the application of biotechnology to plant pathology; the use of alternative crop production methods and the management of diseases that help to protect the environment.

Table of Contents

Chapter 1

Disease Cycle, Development and Management of Sclerotinia Stem Rot of Potato

Dennis A. Johnson[1], Zahi K. Atallah[2]

[1]Department of Plant Pathology, Washington State University, Pullman, USA

[2]Hartnell College, Salinas, USA

ABSTRACT

Potato is severely affected by Sclerotinia sclerotiorum in the US Pacific Northwest (PNW) and satisfactory disease management has not been achieved until recently. Recent research has given a better understanding of key epidemiological factors which has resulted in improved disease management in the PNW. This work reviews the epidemiology of Sclerotinia stem rot and highlights information that has led to the better management of the disease on potato. The primary source of inoculum of Sclerotinia sclerotiorum for potato in the Columbia Basin of the PNW are ascospores produced within potato fields or carried by wind currents from neighboring fields planted to winter cereals or other crops. Ascospores are ejected from apothecia and disseminated throughout the Columbia Basin over an extended time period. Immature apothecia generally first emerge at or shortly after row closure in potato fields in the Columbia Basin. Ascospores of S. sclerotiorum are incapable of direct infection of intact green potato tissues, and flower blossoms are crucial for infection and development of the disease in potato. Airborne ascospores are deposited on open potato blossoms still attached to the canopy. Infested flowers fall and are trapped on stems, usually leaf axils, or fall on the ground, and fungal mycelia then rapidly colonize the blossoms when humidity is high in the plant canopy. Ascospores are also deposited on senescent and dead plant material on the ground, germinate, and produce mycelium. Infection occurs shortly after contaminated blossoms become lodged on stems in the plant canopy, or after stems come in contact with contaminated fallen blossoms or decomposing plant tissues on the ground. Infection can occur within 3 days after contact of contaminated blossoms with green tissue. Lesions initially appear on potato stems 14 to 30 days after row closure and 12 to 20 days following full bloom of primary inflorescences of potato in the Columbia Basin. Significant levels of outcrossing sexual recombination have been found in the homothallic S. sclerotiorum population in the Pacific Northwest. Sclerotinia stem rot of potato is best managed by using an integrated approach of combining cultural practices that produce optimum plant foliar development without producing an excessive crop canopy, irrigation management to avoid excessive amounts of irrigation water, and timely applications of fungicides. Contaminated flower blossoms served as a bridge for infection and fungicides applied before contaminated blossoms drop on foliage most effectively reduce infections on stems. Improved disease management has been achieved by timing fungicides to coincide with full bloom of primary inflorescences.

KEYWORDS

White Mold, Sclerotinia Sclerotiorum, Epidemiology and Fungicide Application Timing

1. INTRODUCTION

Sclerotinia sclerotiorum (Lib.) de Bary, the cause of Sclerotinia stem rot or white mold of potato, is a necrotrophic, inoperculate discomycete of global distribution. The fungus is among the most non-specific of plant pathogens infecting mostly herbaceous plants of diverse phylogenetic backgrounds including more than 400 species in 75 plant families. Cultivated crops susceptible to S. sclerotiorum include potato, bean, soybean, pea, carrot, cucurbits, sunflower, lettuce, cucurbits and mustards. Susceptible weeds include chickweed, redroot pigweed, dandelion and wild clover. Grasses and cereals are not affected [1]. The fungus is both soil- and airborne.

Sclerotia serve as survival structures in the soil and ascospores are the means of aerial dispersal. While the disease is prevalent in potato in the US Pacific Northwest (PNW), its impact is more limited and its occurrence is sporadic in potatoes in Midwestern US states. The disease is primarily promoted by an extensive canopy growth which increases humidity and extends leaf wetness within crop canopies. Increased disease incidences is associated with overhead sprinkler irrigation, a non-upright cultivar architecture, high crop density, close row width, continuous plant surface wetness and excess nitrogen fertilization in potato and other crops [2] - [8].

Satisfactory management of Sclerotinia stem rot on potato has not been achieved in the PNW until recently. Research since 2003 has yielded a better understanding of key epidemiological factors which has resulted in improved disease management. This work reviews the epidemiology of Sclerotinia stem rot and highlights information that has led to the better management of the disease on potato.

2. DISEASE CYCLE

Sclerotinia sclerotiorum is typified by the lack of conidia and the generation of sclerotia and tan-colored, cup- or funnel-shaped apothecia generally 2 - 10 mm in diameter [9]. Sclerotia form from the agglomeration of hyphae in colonized plant tissues. Eight hyaline, unicellular ascospores are produced per ascus.

The first symptoms on potato appear as watersoaked spots usually at stem and branch axils or on branches or stems in contact with the soil. A white cottony growth of fungus mycelium develops on the lesion, and the infected tissue becomes soft and watery. Lesions often expand in size rapidly following establishment and may girdle the stem which causes the foliage to wilt. Lesions become dry and will turn beige, tan or bleached white in color and papery in appearance. Hard, irregularly shaped sclerotia form in and on decaying plant tissues. They are generally 2 - 4 mm in diameter and up to 25 mm long, initially white to cream in color but become black with age, and are frequently found in the hollowed-out center of infected stems. Sclerotia eventually fall to the ground, and are incorporated into the soil during tillage.

Sclerotia are melanized resistant structures capable of withstanding adverse environmental conditions including pesticides and enable the fungus to survive from season to season. They can persist at least 3 years buried in the soil in the absence of a favoring environment for germination. Sclerotia act as a means of pathogen dispersal within fields and over long distances. Because of their size, shape and color, sclerotia are commonly disseminated with the seed of a large number of crops, such as bean or pea. Additionally, the transfer of soil, tillage, agricultural equipment and even water can disseminate sclerotia.

Sclerotia require a conditioning period of cool temperature before germination, and either germinates directly by producing mycelium or carpogenically by producing apothecia [10]. Only the sclerotia in approximately the top 5 cm of the soil are biologically active, sclerotia at lower depths remain dormant. An accumulation of thermal degree days and moist soil, but not water-saturated soil, are required for germination [11] [12]. Direct germination usually occurs before row closure and the mycelium infects the crown area of potato plants in the immediate vicinity, usually a distance of less than 1.5 cm. Apothecia emerge from shallowly buried sclerotia in shaded and moist soils under plant canopies [13] [14]. Multiple apothecia may emerge from a single sclerotium. Apothecia remain functional for 5 - 10 days under shaded conditions in the field and can produce up to 30 million ascospores [14]. Infection of potato is mostly indirectly through ascospores, as is the case with most hosts [11] [15] [16], and apothecial production through carpogenic germination of sclerotia is a fundamental part of the pathogen's disease cycle.

Ascospores are forcibly ejected in mass from the asci in visible puffs of tens of thousands of ascospores. Decreases in relative humidity close to the apothecia are credited for the "puffing" of ascospores [14] [17]. Ascospore releases usually peak around midday and plummet late in the afternoon and at night [4] [18]. Apothecial production and ascospore releases may be staggered over an extended period of several weeks, especially given that soil shading and availability of moisture are required for carpogenic germination. Ascospores are deposited in potato fields in the Columbia Basin from May to mid-July, which precedes row closure by several weeks and extends beyond full bloom of primary flower inflorescences. The number of airborne ascospores decline after full bloom, but increase again in the fall when canola and other mustards are large enough to shade the soil [19].

Ascospores of S. sclerotiorum are associated with short and medium range dispersal. Aerial dispersal of ascospores can reach up to 3 - 4 kilometers, but up to 90% of ascospores are thought to be deposited within 100 - 150 meters from the source [14] [20] - [22]. Viable ascospores have been captured at 6000 m in altitude [23], which indicates a potential dissemination over long distances. Non-germinated ascospores can survive up to 12 days in the crop canopy depending on location in the canopy and environmental conditions. Increasing temperatures above 21 C and exposure to ultraviolet radiation increase ascospore mortality [24]. White mold epidemics in bean and rapeseed occurring early in the season have been traced back to ascospores produced by apothecia in neighboring fields cropped to winter crops [23] [25] [26].

The primary source of inoculum for potato in the Columbia Basin are ascospores produced within potato fields or carried by wind currents from neighboring fields planted to winter cereals or other crops. Fields adjacent to current-season potatoes are generally in rotation with potato or other crops susceptible to S. sclerotiorum and may contain numerous sclerotia. Apothecia are produced in these fields as well as in association with weed hosts in non-cropped land areas adjacent to potato fields. Timing of apothecia development varies among fields and depends on degree of shade due to crop development, soil moisture and temperature [11] [12]. Immature apothecia generally first emerge at or shortly after row closure in potato fields in the Columbia Basin.

Ascospores of S. sclerotiorum are incapable of direct infection of intact green potato tissues, but they must first colonize flowers or senescing or dead plant tissues and then use them as an energy source to infect green tissues [5] [14] [26]. Blossoms are crucial for infection and development of Sclerotinia stem rot in potato. Airborne ascospores are deposited on open potato blossoms still attached to the canopy [19]. Infested flowers fall and are trapped on stems, usually leaf axils, or fall on the ground, and fungal mycelia then rapidly colonize the blossoms when humidity is high in the plant canopy. Ascospores are also deposited on senescent and dead plant material on the ground, germinate, and produce mycelium. Infection occurs shortly after contaminated blossoms become lodged on stems in the plant canopy, or after stems come in contact with contaminated fallen blossoms or decomposing plant tissues on the ground (Figure 1). Secondary spread to additional stems may occur when green stems lodge or lean onto actively expanding lesions on infected stems. Flowers and other senescing tissues colonized by ascospores are generally the cause of the vast majority of lesions as opposed to infections near the soil line initiated by mycelium from soil-borne sclerotia [19].

Flower removal before blossom drop drastically reduced disease incidence compared to potato plants in control plots where blossoms were not removed and gives further evidence of the role of flower blossoms as a bridge for infection (Atallah and Johnson, 2004). Similarly, presence of ascospores on flower petals of canola was positively correlated with disease occurrence [27] and bean blossoms were described as "essential intermediaries" between the fungus and the host [13] [25] [26]. Flower contamination by S. sclerotiorum ascospores was associated with disease incidence in bean and canola, where ascospores remained inactive until blossom senescence, which was followed by a rapid colonization of blossoms and subsequent disease when humidity was favorable [28] -[30].

Occasionally, blighted inflorescences with white mycelium of S. sclerotiorum growing from pedicel and peduncle tissues can be observed attached to potato plants in sprinkler irrigated fields in the Columbia Basin of Washington and Oregon. Remnants of the senesced blossom may still be attached to the inflorescence. Evidently ascospores were deposited and then germinated in the flower and the fungus then grew past the abscission ring before the blossom dropped (Figure 2). This illustrates the potential role of contaminated blossoms as an infection bridge from a saprophytic to parasitic phase, especially when realizing that most contaminated blossoms abscise and drop into the canopy or to the ground.

Immature apothecia generally first emerge at or shortly after row closure in potato fields in the Columbia Basin. Apothecia are not found after diligent searching in some potato fields, but Sclerotinia stem rot may still develops in many of these fields [19]. Sclerotinia stem rot has not been observed in a few locations in the Columbia Basin were potato fields are isolated from susceptible annual crops by perennial orchards, the Columbia River, and/or non-agricultural land areas [31]. Apothecia present in the potato field, in neighboring potato fields, or in fields of other crops in rotation with potatoes or crops susceptible to S. sclerotiorum are likely sources of ascospore inoculum. Ascospores originating external to a potato field appear to be an important and abundant source of inoculum. Distance of dissemination is not known, but is likely fairly limited based on the large number of susceptible crops grown in the Columbia Basin which may have previously produced sclerotia. Certainly, the closer the inoculum source, including apothecia within a potato field, the greater the number of ascospores impacting and contaminating potato blossoms.

Ascospores are ejected from apothecia and disseminated throughout the Columbia Basin over an extended time period. The greatest numbers of ascospores of S. sclerotiorum were captured after row closure and about a week before and continuing a few days after full bloom of primary inflorescences. A hundred percent of the blossoms collected at full bloom of primary inflorescences from potato plants in eight of ten fields were contaminated with S. sclerotiorum and initial stem infections originating from airborne ascospores and the subsequent mycelial growth on an organic substrate were not observed until after blossom drop in the Columbia Basin [19].

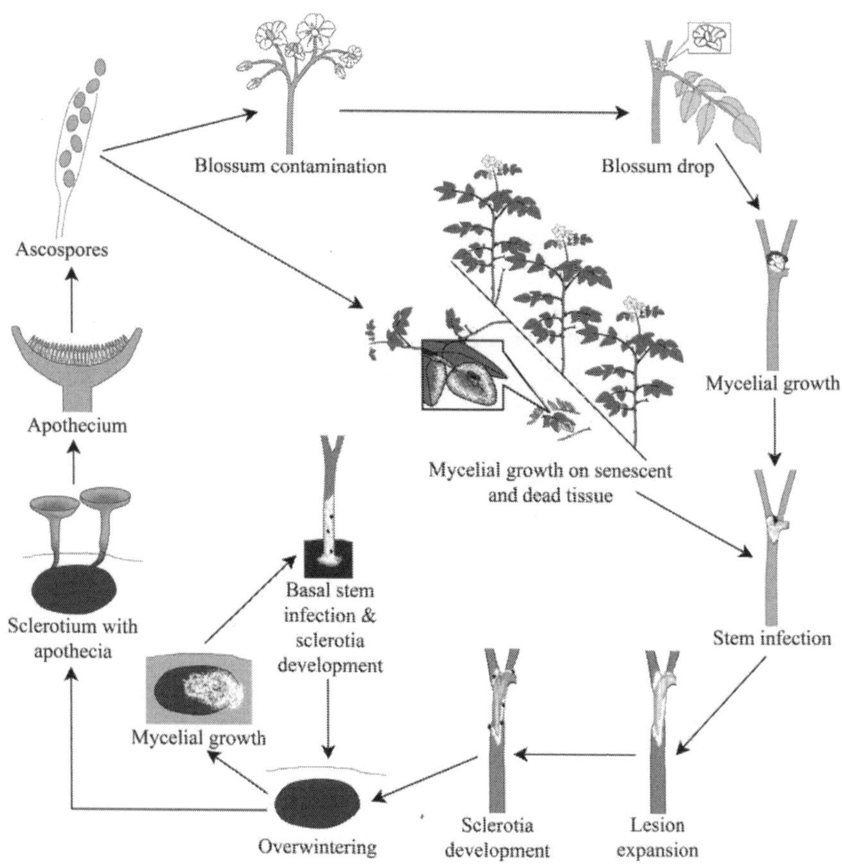

Figure 1. Disease cycle of Sclerotinia sclerotiorum on potato.

In an experiment in the Columbia Basin, Petri plates of a semi-selective medium for S. sclerotiorum were exposed to capture ascospores at 30 - 40 m intervals along a straight-line transected from 60 m outside of potato fields to 80 m into fields on opposite sides of six commercial center pivot irrigated fields over two years. Fields were at least 51 ha in area, with a minimum radius of 403 m. Ascospores were captured outside of the potato fields, at the edge of the fields and in the fields. A gradient in numbers of ascospores deposited in Petri dishes over distance was not evident. Equal numbers of ascospores were captured in the windward and leeward sides of the six circles [31]. Ascospores were captured in fields before row closure, which was before apothecia were present in fields, and they were captured in fields where apothecia were not observed to develop in two studies [19] [31]. From these experiments, it is evident that in the Columbia Basin, relatively large numbers of ascospores originate outside as well as within potato fields. Because of the major role that external ascospores play in the infection of potato, sanitation practices to reduce sclerotia in soil, such as a biological control agent or and soil fumigation, likely will not adequately control Sclerotinia stem rot of potato in the Columbia Basin. Extrinsic sources of ascospore inoculum will likely overcome the effects of sanitation within the field. However, when large numbers of viable sclerotia are present in a field usually due to tight rotations with susceptible crops, reducing the numbers of these sclerotia by crop rotation or biological agents may be beneficial in reducing numbers of basal stem infections and numbers of apothecia. Extrinsic sources of ascospores have been shown to be important for epidemics of white mold in bean [32], lettuce drop [33] and Sclerotinia stem rot in rapeseed [34].

3. DISEASE DEVELOPMENT

Mycelial mats of S. sclerotiorum are usually first observed colonizing senesced blossoms that have dropped on stems in the plant canopy and on the ground 2 to 3 days after blossoms begin dropping from primary flower inflorescences. Infection can occur within 3 days after contact of contaminated blossoms with green tissue. Lesions initially appear on potato stems 14 to 30 days after row closure and 12 to 20 days following full bloom of primary inflorescences of potato in the Columbia Basin (Figure 3) [19] [35]. Time period between row closure and full bloom of primary inflorescences varies depending on cultivar and environment; full bloom of primary inflorescences occurred 5 to 10 days after row closure over a five year period for cultivars Shepody and Ranger Russet in the Columbia Basin. Senescent and dead plant materials on the canopy floor that potentially serve as saprophytic substrates for mycelial mats of S. sclerotiorum are usually scarce prior to row closure, as restricted sun light in the lower canopy appears to be associated with yellowing and senescence of lower leaves. At blossom drop large amounts of senescent and dead plant material may be available and can support numerous mycelial mats on the ground. Potato stems that become lodged and cultivars that have a prostate habit can be severely infected at this time.

Figure 2. Blighted inflorescences with white mycelium of S. sclerotiorum growing from pedicel and peduncle tissues.

Lesions on stems rapidly expand over a period of 4 to 10 days after initial infection. Disease progress curves for incidence of infected stems in fields also increase rapidly over a period of 2 to 3 weeks after symptoms initially appear and then usually levels off [19]. The shape of the disease progress curves are consistent with a favoring environment for infection and inoculum arriving on the host as ascospore-contaminated blossoms during blossom drop. The extent of additional infections depends on numbers of stems becoming prostrate on the ground, due to cultivar habit or environment and on frequency of sprinkler irrigation water and rainfall. Sclerotinia stem rot lesions then become less active as the potato canopy begins to open and as less irrigation water is applied due to decreasing plant needs.

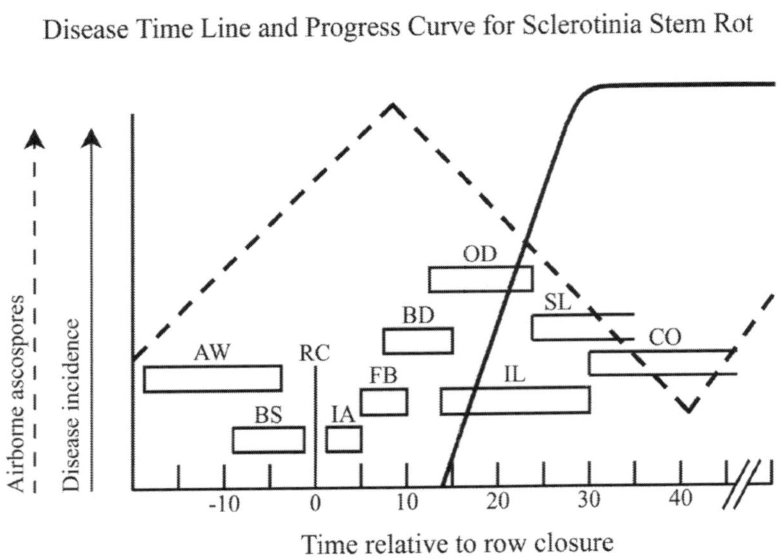

AW = Apothecia in winter cereals and other crops
BS = Basal stem lesions from directly germinating (mycelium) sclerotia
RC = Row closure
IA = Initial apothecia in potato field
FB = Full bloom of primary inflorescences
BD = Blossum drop of primary inflorescences
OD = Abundant organic debris on ground; initial mycelial mats on ground
IL = Initial lesions
SL = Stems lodge on canopy floor
CO = Canopy opens and lesions become inactive

Figure 3. Disease time line and progress curve of Sclerotinia stem rot on potato cultivars Ranger Russet and Shepody in the Columbia Basin of Washington.

The carpogenic germination of sclerotia leading to the production of apothecia and subsequent release and dissemination of ascospores is the result of a sexual recombination. Clonal populations of S. sclerotiorum were repeatedly encountered across 2000 Km in the Canadian Great Plains and the fungus is thought to be primarily homothallic in agricultural areas [36] - [38]. Nevertheless, significant levels of outcrossing sexual recombination have been found to occur in the Pacific Northwest (PNW) and Australia [39] [40]. Additional evidence of outcrossing in populations of the homothallic S. sclerotiorum collected from agricultural fields has been demonstrated using linkage disequilibrium decay [41]. Previously, evidence for outcrossing under non-agricultural conditions has been noted [42], while research results have hinted for possible, yet infrequent outcrossing in agricultural conditions [42] [43]. Successful forced outcrossing by Ford et al. (1995) indicated that mycelial compatibility (MCG) and vegetative compatibility (VCG) are governed by different genetic loci. Outcrossing could interfere with the development of stem rot resistant potato cultivars and lead to the distribution of agriculturally disadvantageous traits, such as increased pathogen aggressiveness, as well as resistance to specific fungicides. In line with differing sexual lifestyles between the Canadian Great Plains and the PNW, differences in aggressiveness among isolates were observed in Canada [44], while no differences were observed in the PNW [39].

4. DISEASE MANAGEMENT

Sclerotinia stem rot of potato is best managed by using an integrated approach of combining cultural practices that produce optimum plant foliar development without producing an excessive crop canopy, irrigation management to avoid excessive amounts of irrigation water, and timely applications of fungicides. Potato cultivars with an erect habit usually escape infections from mycelial mats on the soil surface; whereas, cultivars with a prostate habit can have high disease incidence.

Resistance—Potato cultivars are not available with resistance to infection. Disease severity can be influenced by cultivar growth habit. For example, cultivars Russet Ranger, Russet Burbank, Russet Norkotah, Shepody, Gem Russet, Atlantic, Alturas Russet, and Umatilla Russet were equally susceptible to Sclerotinia stem rot in controlled, replicated, inoculation experiments (unpublished data). Potato cultivar Gem Russet has a prostate habit and extremely high incidences of Sclerotinia stem rot may occur on this cultivar in the field. With an absence of resistant cultivars, fungicide applications coupled with judicious irrigation management remain the most effective management method.

Fungicide application timing—Control of Sclerotinia stem rot was significantly improved when fungicides thiophanate-methyl (Topsin M 70 WP) at 1.176 kg ai/ha, fluazinam (Omega 500F) at 0.280 kg ai/ha, and boscalid (Endura) at 0.336 kg ai/ha were applied at full bloom of primary inflorescences compared to applications at closure between rows (row closure) in cultivars Shepody (determinate) and Ranger Russet (indeterminate). Contaminated flower blossoms served as a bridge for infection and fungicides applied before contaminated blossoms dropped on foliage subsequently reduced infections on stems [35]. Similar observations on timing of fungicide application were made when bean flowers were protected with benomyl at full bloom, which prevented white mold development even under optimal environmental conditions [14] [26]. In those studies, benomyl applications made after full bloom of bean failed to provide effective stem rot control because flowers had already acquired the inoculum.

A protective cover of fungicide on potato stems and on potential substrates for saprophytic growth of S. sclerotiorum may both be needed for most effective chemical control. Fungicides applied before full bloom of primary inflorescences are possibly washed from stems and partially degraded before inoculum on colonized blossoms and other substrates is present; whereas, applications at full bloom of primary inflorescences are made just before infection would otherwise occur from fallen contaminated blossoms. Colonization periods (time from inoculation with colonized blossoms to initial stem lesions on potato stems) of three days have been observed in greenhouse experiments. This is consistent with the rapid development of lesions that is often observed soon after contaminated blossoms drop in a humid potato canopy. Initial fungicide application should be made at full bloom of primary inflorescences and not at row closure. This generally occurs 5 to 10 days after row closure in the Columbia Basin. In cultivars that flower over an extended period such as Russet Burbank, initial fungicide application in the Columbia Basin has been most effective about 7 days after row closure.

A difference in application timing of a few days made a difference in the effectiveness of the fungicides. Fungicides applied at row closure did not cover the flower blossoms because blossoms were absent at that time, and fungicides applied at 20% blossom drop were too late to effectively prevent stem infections from infested blossoms. Data collected two of three years indicated that the window for effective applications was somewhat wider than just at 100% bloom. Nevertheless, the width of window for the most effective control truncated before 20% drop of primary flower inflorescences [35].

Applications of boscalid, fluazinam and tiophanate-methyl at full bloom of primary inflorescences, did not vary in efficacy and all three fungicides effectively reduced Sclerotinia stem rot. Thiophanate-methyl was generally less effective than either fluazinam or boscalid in reducing number of stems with lesions when applied at row closure and 10% bloom of primary inflorescences in two of three years [35].

Foliar fungicides currently registered for Sclerotinia stem rot would not be expected to prevent crown infections from directly germinating sclerotia. This is because the inoculum exists in or on the soil, and soil around the lower stems prevents contact of the fungicide with the stem.

Cultural practices—Severity of Sclerotinia stem rot can be reduced with a combination of practices that limiting potato vine growth such as nitrogen fertilizer management and avoiding over irrigation. Cultural practices need to be employed before infections begin in a field. Irrigation practices that promote long periods of leaf wetness or high relative humidity within the crop canopy should be avoided. Irrigation should be restricted during rainy weather, and on cool, cloudy days, whenever possible.

Biological control—The use of the biological control agent Coniothyrium minitans, a mycoparasite that feeds on sclerotia of S. sclerotiorum, to reduce the sclerotia bank in the soil has yielded conflicting results between the regions where experiments have been conducted. In Wisconsin, applications of C. minitans have repeatedly reduced white mold incidence in bean fields (W. R. Stevenson, personal communication). In addition, application of C. minitans to soil at seeding time reduced apothecial product from sclerotia under the canopies of bean, canola, wheat, and barley in Alberta [45]. Conversely, numbers of apothecia, number of blossom infections, and incidence of stem rot infections were not reduced by applications of C. minitans compared to the non-treated control in large, replicated potato plots in a study in the Columbia Basin of Washington State [31]. In the cited study, a commercial preparation of C. minitans was applied by chemigation just after the last soil tillage operation in six fields over two years. The application timing was 3 to 5 weeks before row closure in the six fields which should have been sufficient time for the mycoparasite to reduce sclerotia numbers. In addition to the lack of control from not reducing number of sclerotia and subsequent apothecia in the treated plots, the mycoparasite would not prevent airborne ascospores from migrating from non-treated, neighboring and more distant fields. This would likely undo any sanitation effects on sclerotia in the treated field and appears to be more substantial in the PNW than Wisconsin. Additionally, deep tillage and hilling of fields will re-distribute sclerotia across soil profiles, thus bringing to the top sclerotia that have not been exposed to the biological control agent.

5. FUTURE RESEARCH

Additional improvements in disease management of Sclerotinia stem rot in potato are needed to decrease the expense of fungicides. Questions needing answers to improve control include: What is the quantitative impact of non-blossom, senescing and dead tissues on disease development relative to blossoms? What is the role and prevalence of stem infections initiated by S. sclerotiorum on possible subsequent development of aerial stem rot caused by soft rot bacteria? Does S. sclerotiorum infection cause a significant number of entry sites for the bacteria? What is the effective dissemination range for ascospores of S. sclerotiorum? What is the role of protectant fungicides in reducing disease incidence—do they prevent colonization of blossoms, protect green stems from infection by colonized blossoms or both?

Timing of initial fungicide application is based on crop phenology in the Columbia Basin, but development and implementation of disease forecasting models or warning systems would be useful in identifying fields that need or do not need to be treated with fungicide, determine the need and schedule for additional applications, and identify fields and areas that may potentially benefits from application of a biological agent. Potentially useful forecasting models have been developed [11] [12] [27].

Lesion expansion is an important epidemic component for Sclerotinia stem rot [46]. A greater understanding is needed on the quantitative effects of environmental factors such as leaf wetness, humidity, and temperature on rates of lesion expansion on potato genotypes.

Additionally, modifying potato cultivars to provide an upright growth habit and a more aerated canopy should reduce severity of Sclerotinia stem rot in potato as it has in other crops including bean, soybean, canola and peanut

[29] [47] -[49]. In commercial potato production, true seed is not used for propagation, yet flowers are consistently produced and fruits are often formed in fields during certain growing seasons. A flowerless cultivar has less disease pressure than flowering cultivars in the PNW and researchers working on identifying sources and mechanisms of resistance to S. sclerotiorum may find that reducing blooming to be the alternative to the so-far elusive resistance in potato.

ACKNOWLEDGEMENTS

Dennis A. Johnson, Zahi K. Atallah "PPNS # 0669", Department of Plant Pathology, College of Agricultural, Human, and Natural Resource Sciences, Agricultural Research Center, Hatch Project No. WNP0678, Washington State University, Pullman, WA 99164-6430, USA.

REFERENCES

1. Boland, G.J. and Hall, R. (1994) Index of Plant Hosts of Sclerotinia sclerotiorum. Canadian Journal of Plant Pathology, 16, 93-108.

2. Grau, C.R. and Radke, V.L. (1984) Effects of Cultivars and Cultural-Practices on Sclerotinia Stem Rot of Soybean. Plant Disease, 68, 56-58.

3. Grogan, R.G. and Abawi, G.S. (1975) Influence of Water Potential on Growth and Survival of Whetzelinia sclerotiorum. Phytopathology, 65, 122-128.

4. Gutierrez, W.A. and Shew, H.D. (1998) Identification and Quantification of Ascospores as the Primary Inoculum for Collar Rot of Greenhouse-Produced Tobacco Seedlings. Plant Disease, 82, 485-490.

5. Keay, M.A. (1939) A Study of Certain Species of the Genus Sclerotinia. Annals of Applied Biology, 26, 227-246.

6. Natti, J.J. (1971) Epidemiology and Control of Bean White Mold. Phytopathology, 61, 669-674.

7. Purdy, L.H. (1979) Sclerotinia Sclerotiorum—History, Diseases and Symptomatology, Host Range, Geographic Distribution, and Impact. Phytopathology, 69, 875-880.

8. Steadman, J.R. (1979) Control of Plant-Diseases Caused by Sclerotinia Species. Phytopathology, 69, 904-907.

9. Kohn, L.M. (1979) Delimitation of the Economically Important Plant Pathogenic Sclerotinia Species. Phytopathology, 69, 881-886.

10. Dillard, H.R., Ludwig, J.W. and Hunter, J.E. (1995) Conditioning Sclerotia of Sclerotinia sclerotiorum for Carpogenic Germination. Plant Disease, 79, 411-415.

11. Clarkson, J.P., Phelps, K., Whipps, J.M., Young, C.S., Smith, J.A. and Watling, M. (2004) Forecasting Sclerotinia Disease on Lettuce: Toward Developing a Prediction Model for Carpogenic Germination of Sclerotia. Phytopathology, 94, 268-279.

12. Sun, P. and Yang, X.B. (2000) Light, Temperature, and Moisture Effects on Apothecium Production of Sclerotinia sclerotiorum. Plant Disease, 84, 1287-1293.

13. Schwartz, H.F. and Steadman, J.R. (1978) Factors Affecting Sclerotium Populations of, and Apothecium Production by Sclerotinia sclerotiorum. Phytopathology, 68, 383-388.

14. Steadman, J.R. (1983) White Mold—A Serious Yield-Limiting Disease of Bean. Plant Disease, 67, 346-350.

15. Phillips, A.J.L. (1987) Carpogenic Germination of Sclerotia of Sclerotinia sclerotiorum: A Review. Phytophylactica, 19, 279-283.

16. Ben-Yephet, Y., Genizi, A. and Siti, E. (1993) Sclerotial Survival and Apothecial Production by Sclerotinia sclerotiorum Following Outbreaks of Lettuce Drop. Phytopathology, 83, 509-513.

17. McCartney, H.A. and Lacey, M.E. (1991) The Relationship between the Release of Ascospores of Sclerotinia sclerotiorum, Infection and Disease in Sunflower Plots in the United Kingdom. Grana, 30, 486-492.

18. McCartney, H.A. and Lacey, M.E. (1992) Release and Dispersal of Sclerotinia Ascospores in Relation to Infection. Brighton Crop Protection Conference: Pests and Diseases, 3A-2, 109-116.

19. Atallah, Z.K. and Johnson, D.A. (2004) Development of Sclerotinia Stem Rot in Potato Fields in South-Central Washington. Plant Disease, 88, 419-423.

20. Abawi, G.S. and Grogan, R.G. (1979) Epidemiology of Diseases Caused by Sclerotinia Species. Phytopathology, 69, 899-904.

21. Ben-Yephet, Y. and Bitton, S. (1985) Use of a Selective Medium to Study the Dispersal of Ascospores of Sclerotinia sclerotiorum. Phytoparasitica, 13, 33-40.

22. Stelfox, D., Williams, J.R., Soehngen, U. and Topping, R.C. (1978) Transport of Sclerotinia sclerotiorum Ascospores by Rapeseed Pollen in Alberta. Plant Disease Reporter, 62, 576-579.

23. Williams, J.R. and Stelfox, D. (1979) Dispersal of Ascospores of Sclerotinia sclerotiorum in Relation to Sclerotinia Stem Rot of Rapeseed. Plant Disease Reporter, 63, 395-399.

24. Caesar, A.J. and Pearson, R.C. (1983) Environmental Factors Affecting Survival of Ascospores of Sclerotinia sclerotiorum. Phytopathology, 73, 1024-1030.

25. Abawi, G.S. and Grogan, R.G. (1974) Ascospores of Whetzelinia sclerotiorum as Primary Inoculum Causing White Mold of Beans n New York. Phytopathology, 64, 578-578.

26. Abawi, G.S., Polach, F.J. and Molin, W.T. (1975) Infection of Bean by Ascospores of Whetzelinia sclerotiorum. Phytopathology, 65, 673-678.

27. Turkington, T.K. and Morrall, R.A.A. (1993) Use of Petal Infestation to Forecast Sclerotinia Stem Rot of Canola: The Influence of Inoculum Variation over the Flowering Period and Canopy Density. Phytopathology, 83, 682-689.

28. Boland, G.J. and Hall, R. (1988) Epidemiology of Sclerotinia Stem Rot of Soybean in Ontario. Phytopathology, 78, 1241-1245.

29. Coyne, D.P., Steadman, J.R. and Anderson, F.N. (1974) Effect of Modified Plant Architecture of Great Northern Dry Bean Varieties (Phaseolus vulgaris) on White Mold Severity, and Components of Yield. Plant Disease Reporter, 58, 379-382.

30. Lefol, C. and Morrall, R.A.A. (1996) Immunofluorescent Staining of Sclerotinia Ascospores on Canola Petals. Canadian Journal of Plant Pathology, 18, 237-241.

31. Hammond, C. and Johnson, D.A. (2005) Origin of Ascospores of Sclerotinia sclerotiorum for White Mold Development in Potato Fields in the Columbia Basin. American Journal of Potato Research, 82, 72.

32. Boland, G.J. and Hall, R. (1987) Epidemiology of White Mold of White Bean in Ontario. Canadian Journal of Plant Pathology, 9, 218-224.

33. Patterson, C.L. and Gorgan, R.G. (1985) Differences in Epidemiology and Control of Lettuce Drop Caused by Sclerotinia minor and S. sclerotiorum. Plant Disease, 69, 766-770.

34. Morrall, R.A.A. and Dueck, J. (1982) Epidemiology of Sclerotinia Stem Rot of Rapeseed in Saskatchewan. Canadian Journal of Plant Pathology, 4, 161-168.

35. Johnson, D.A. and Atallah, Z.K. (2006) Timing Fungicide Applications for Managing Sclerotinia Stem Rot of Potato. Plant Disease, 90, 755-758.

36. Hambleton, S., Walker, C. and Kohn, L.M. (2002) Clonal Lineages of Sclerotinia sclerotiorum Previously Known from Other Crops Predominate in 1999-2000 Samples from Ontario and Quebec Soybean. Canadian Journal of Plant Pathology, 24, 309-315.

37. Kohli, Y. and Kohn, L.M. (1998) Random Association among Alleles in Clonal Populations of Sclerotinia sclerotiorum. Fungal Genetics and Biology, 23, 139-149.

38. Kohli, Y., Morrall, R.A.A., Anderson, J.B. and Kohn, L.M. (1992) Local and Trans-Canadian Clonal Distribution of Sclerotinia sclerotiorum on Canola. Phytopathology, 82, 875-880.

39. Atallah, Z.K., Larget, B., Chen, X. and Johnson, D.A. (2004) High Genetic Diversity, Phenotypic Uniformity, and Evidence of Outcrossing in

Sclerotinia sclerotiorum in the Columbia Basin of Washington State. Phytopathology, 94, 737-742.

40. Sexton, A.C. and Howlett, B.J. (2004) Microsatellite Markers Reveal Genetic Differentiation among Populations of Sclerotinia sclerotiorum from Australian Canola Fields. Current Genetics, 46, 357-365.

41. Attanayake, R.N., Tennekoon, V., Johnson, D.A., Porter, L.D., del Rio-Mendoza, L., Jiang, D. and Chen, W. (2014) Inferring Outcrossing in the Homothallic Fungus Sclerotinia sclerotiorum Using Linkage Disequilibrium Decay. Heredity, 113, 353-363.

42. Kohn, L.M. (1995) The Clonal Dynamic in Wild and Agricultural Plant—Pathogen Populations. Canadian Journal of Botany, 73, 1231-1240.

43. Cubeta, M.A., Cody, B.R., Kohli, Y. and Kohn, L.M. (1997) Clonality in Sclerotinia sclerotiorum on Infected Cabbage in Eastern North Carolina. Phytopathology, 87, 1000-1004.

44. Kohli, Y., Brunner, L.J., Yoell, H., Milgroom, M.G., Anderson, J.B., Morrall, R.A.A. and Kohn, L.M. (1995) Clonal Dispersal and Spatial Mixing in Populations of the Plant Pathogenic Fungus Sclerotinia sclerotiorum. Molecular Ecology, 4, 69-77.

45. McLaren, D.L., Huang, H.C. and Rimmer, S.R. (1996) Control of Apothecial Production of Sclerotinia sclerotiorum by Coniothyrium minitans and Talaromyces flavus. Plant Disease, 80, 1373-1378.

46. Berger, R.D., Bergamin Filho, A. and Amorim, L. (1997) Lesion Expansion as an Epidemic Component. Phytopathology, 87, 1005-1013.

47. Butzler, T.M., Bailey, J. and Beute, M.K. (1998) Integrated Management of Sclerotinia Blight in Peanut: Utilizing Canopy Morphology, Mechanical Pruning, and Fungicide Timing. Plant Disease, 82, 1312-1318.

48. Fuller, P.A., Steadman, J.R. and Coyne, D.P. (1984) Enhancement of White Mold Avoidance and Yield in Dry Beans by Canopy Elevation. HortScience, 19, 78-79.

49. Steadman, J.R., Coyne, D.P. and Cook, G.E. (1973) Reduction of Severity of White Mold Disease on Great-Northern Beans by Wider Row Spacing and Determinate Plant-Growth Habit. Plant Disease Reporter, 57, 1070-1071.

Chapter 2

Use of Biofungicides for Controlling Plant Diseases to Improve Food Availability

Antonieta De Cal [1], Inmaculada Larena [1], Belén Guijarro [2] and Paloma Melgarejo [1,*]

[1]Department of Plant Protection, INIA, Carretera de La Coruña km 7. 28040 Madrid, Spain

[2]Department of Botany, Biological Sciences 2, La Trobe University, Bundoora, Vic. 3086, Australia

ABSTRACT

Biological control of fungal plant pathogens can improve global food availability, one of the three pillars of food security, by reducing crop losses, particularly for low-income farmers. However, the interrelationships of many environmental variables can result in multiple interactions among the organisms and their environment, several of which might contribute to effective biological control. Here, we present an advanced survey of the nature and practice of biological control when it is used to control brown rot in stone fruit. Specifically, we describe the population dynamics of *Penicillium frequentans* and *Epicoccum nigrum* and their efficacy as biocontrol agents against brown rot disease under field conditions. The size of *P. frequentans* population after an application of a *P. frequentans* conidial formulation during the crop season is bigger than that of *E. nigrum* following the application of an *E. nigrum* conidial formulation. Moreover, applications of a *P. frequentans* conidial formulation during the crop season also caused a higher reduction in the number of *Monilinia* spp. conidia on the fruit surface than that found after applications of an *E. nigrum*formulation during the growing season.

KEYWORDS

Penicillium frequentans; *Epicoccum nigrum*; brown rot; stone fruit; population dynamics; biocontrol

1. INTRODUCTION

Plant diseases need to be controlled in order to maintain global food availability, one of the three pillars of food security, and different approaches can be used to prevent, mitigate, or control plant diseases. In addition to good agronomic and horticultural practices, growers can apply chemical fertilizers and pesticides to their crops, and such applications have contributed significantly to spectacular improvements in food security. Finding non-chemical strategies for controlling crop diseases is of considerable interest because of environmental and health concerns about the widespread use of synthetic pesticides [1]. Nowadays, strict regulations on chemical pesticide use exist, and the political pressure to remove the most hazardous chemicals from the market is increasing. Therefore, the use of synthetic pesticides will probably be progressively reduced in the future, and this reduction will be accompanied by an increased reliance on the use of microorganisms as antagonists of plant pathogens [1]. However, controlling the wide spectrum of pathogens using microorganisms still remains an unfulfilled goal in commercial agriculture.

The main shortcoming of microbiological control is its lack of reliability, and several reasons can account for this limitation. Firstly, the population dynamics of the biological antagonist, the diversity of the natural population of the pathogen, and the natural climatic conditions are much more variable than those that are used in the laboratory to study the modes of actions of a biological control agent (BCA). Secondly, the climatic conditions that most favor the development of the pathogen might not be the same as the conditions that are required for the maximum expression of a BCA's antagonistic activities [2]. Therefore, much more research should be devoted to identifying the environmental conditions that are required for the expression of a BCA's beneficial effects under field conditions in order to make microbiological control more reliable.

Brown rot caused by *Monilinia* spp. is an economically important fungal disease of stone fruit, and is responsible for substantial pre- and postharvest losses [3]. Brown rot can be caused by the three *Monilinia* spp., namely *M. laxa, M. fructicola*, and *M. fructigena* [3]. Preharvest rains, the accumulation of excess moisture on fruit, wind, and the movement of insect vectors usually increase the incidence of brown rot [4]. Pre- and postharvest application of fungicides and insecticides to the fruit at proper intervals, the careful handling of harvested fruit in order to avoid wounding, proper pre- and postharvest sanitation, the rapid cooling of the harvested fruit and its storage at 0°C are all currently-used methods for decreasing brown rot incidence [5,6,7]. Despite the use of these disease control methods, brown rot at postharvest can still occur and cause significant crop losses [7,8,9]. Direct yield losses result from the infection of flowers (flower and twig blight), and from fruit rot at the preharvest, harvest, and postharvest stages [10]. Although the fruit can be infected with *Monilinia* spp. at any stage of its development, the disease only becomes severe when the fruit begins to ripen [11,12,13]. Susceptibility of peach fruit to *M. laxa* infection is usually greatest between the 32nd and 34th week of growth when the endocarp and pericarp are completely formed [13]. Postharvest losses are typically more severe than preharvest losses [10], and routinely occur during the fruit's storage and transport [8]. Furthermore, development of pathogen resistance to synthetic pesticides [14,15,16,17,18], strong public opinion against the use of synthetic fungicides and insecticides, and the health risks that are associated with their application have made non-agrochemical products undesirable for controlling brown rot disease [19,20].

Epicoccum nigrum Link and *Penicillium frequentans* Westling are two common constituents of the resident mycobiota of peach twigs and flowers [21]. These two fungi have been shown to reduce twig blight caused by *M. laxa* in experimental stone fruit orchards [22,23,24] and postharvest brown rot caused by *M. laxa* and *M. fructigena* in laboratory assays [25,26,27]. Competition and antibiosis are the mode of action of these two fungal antagonists [24,28,29]. The major impediment in the commercialization of biocontrol products is the development of a formulation that will retain the same efficacy that is observed in laboratory assays in commercial orchards [30]. *E. nigrum* and *P. frequentans* have shown good potential for development as a commercial BCA against the brown rot of peach fruit [25,27,31]. Despite taking precautions to enhance BCA survival, severe environmental conditions may drastically limit a BCA's ability to establish itself at its target site on the host [32]. Understanding the underlying ecological mechanisms of disease control and the population dynamics of a BCA in the environment is required when a BCA is proposed as a replacement for the existing methods of disease control [33]. Hence, the effect of inoculum type, the application frequency, and the time of application need to be zealously studied in order to ensure the efficacy of biological control [34,35].

Therefore, this article is an advanced survey of the nature and practice of biological control when it is used to control brown rot in stone fruit. This review will (i) describe the population dynamics of *Penicillium frequentans* and *Epicoccum nigrum* and their efficacy as BCAs against brown rot disease under field conditions, (ii) explore the relationships between antagonistic populations and biological control, and (iii) briefly outline future directions that might lead to the development of more diverse and effective BCAs.

2. RESULTS AND DISCUSSION

P. frequentans and *E. nigrum* are effective biofungicides because they can reduce the number of *Monilinia* spp. conidia on fruit surfaces. The number of *Monilinia* spp. conidia on peach and nectarine fruit surfaces that were treated with either an *E. nigrum* or *P. frequentans* conidial formulation, as measured by the area under the conidial number of *Monilinia* spp. progress curve (AUncMOPC) are significantly lower ($p= 0.05$) than that found on untreated fruit surfaces (Table 1 and Table 2). A positive relationship between the conidial numbers of *Monilinia* spp. on peach surfaces and the incidence of latent infection has been reported [36], and these latent infections are also related to the incidence of postharvest brown rot [12]. The number of *Monilinia* spp. conidia on non-treated nectarine and peach surfaces increases from flowering to harvest. The numbers of *Monilinia* spp. conidia on fruit surfaces also increase rapidly during the pit hardening stage and during the month immediately before harvest, and the levels can be as high as $10^{3.4}$–$10^{4.5}$ conidia per fruit surface. The highest levels of*Monilinia* spp. conidia on the fruit surface are usually found at harvest. The smallest *Monilinia* spp. conidial population was recorded on peach and nectarine fruit surfaces that were treated with a *P. frequentans* conidial formulation, followed by treatment with an *E. nigrum* conidial formulation and a fungicide (Table 1 and Table 2). In a previous study, we showed that an application of fresh *E. nigrum* or *P. frequentans* conidia at bloom and at preharvest reduced the incidence of postharvest brown rot in commercial peach orchards [31,37]. It has also been reported that preharvest application of other fungal BCAs, such as *Metschnikowia fructicola*, can also reduce the incidence of postharvest rot infections in stored strawberries [38]. Smilanick *et al.* [39] reported that preharvest infection of nectarines and peaches by *Monilinia* spp. could be the reason for poor control of brown rot when BCAs against this disease are applied postharvest.

One of the goals of a population dynamics study is to identify recurring patterns in the dynamics of the population, and to acquire knowledge on the mechanisms that generate these patterns [40]. The population dynamics of *E. nigrum* and *P. frequentans* in two commercial orchards in 2004 and 2005 are shown in Figure 1 and Figure 2, respectively. We found that the number of conidia (AUncPC) and the number of colony-forming units (CFUs) (AUcfuPC) for peach and nectarines trees that were treated with a *P. frequentans* conidial formulation were significantly higher ($p = 0.05$) than those for those trees that were treated with an *E. nigrum* conidial formulation (Table 1 and Table 2). In all instances, the number of *E. nigrum* and *P. frequentans* conidia on the peach surfaces (flowers or fruit) at each sampling date in every orchard was 100- to 1000-fold higher than the number of CFUs on Petri dishes that contained potato dextrose agar (PDA). Significantly larger *E. nigrum*and *P. frequentans* populations ($p= 0.05$) were present on the fruit surfaces 15 days before harvest (15–30% for the conidial populations and 80–100% for the number of CFUs) than on the flower surfaces in the early spring (Figure 1 and Figure 2). The number of *E. nigrum* and *P. frequentans* conidia remained fairly constant on the fruit surfaces over the period from petal fall to the first preharvest treatment with the conidial formulations (Figure 1 and Figure 2). Although the conidial concentration in the *E. nigrum* or *P. frequentans* formulations that were applied to the peach trees were the same, the size of the *P. frequentans* population on the surfaces was higher than that of the *E. nigrum* population.

Effective colonization, a large population size, and the viability of a BCA on plant surfaces are all essential for successful biocontrol of plant diseases [41]. We found that the numbers of *E. nigrum* and *P. frequentans* conidia are always significantly higher ($p = 0.05$) immediately following application of their conidial formulations when compared with those found at other sampling times (Figure 1 and Figure 2). It is possible that the greater number of *P. frequentans* conidia on the fruit surfaces may be due to a better adhesive ability than *E. nigrum* conidia, despite the ten-fold larger size of *E. nigrum* conidia [42]. In addition, the cell wall of *E. nigrum* conidia is composed of two layers whose outer layer is thick, studded with wart-like knobs, and pigmented by indole melanin [42].

Table 1. Effect of the *Epicoccum nigrum* (FOR7) and *Penicillium frequentans* (FOR8) conidial formulations on the area under progress curve for the number of *Monilinia*spp. conidia (AUncMOPC) and the number of *E. nigrum* or *P. frequentans* conidia (AUncPC), and the area under progress curve for the number of colony-forming units (CFU) of *E. nigrum* or *P. frequentans* (AUcfuPC) for the surfaces of "Red September" peaches from a commercial orchard in Alfarrás, Spain in 2004 and 2005.

Year	Formulation	AUncMOPC	AUncPC	AUcfuPC
2004	FOR7	149,902 b	3.59×10^6 a	43,873 a
	FOR8	70,801 c	1.55×10^7 b	168,521 b
	cyproconazole	241,333 b	-	-
	control	837,524 a	-	-
2005	FOR7	479,941 b	1.37×10^7 b	152,798 b
	FOR8	49,326 c	1.39×10^7 b	83,866 a
	tebuconazole	408,852 b	-	-
	control	565,004 a	-	-
	MSE	1.96×10^{11}	1.80×10^{13}	4.19×10^9

*Data are the mean of four replicates. Data were analyzed by analysis of variance. Means followed by the same letter in each column were not significantly different according to the Student–Newman–Keuls multiple range test (p=0.05). MSE, mean squared error.

Table 2. Effect of the *Epicoccum nigrum* (FOR7) and *Penicillium frequentans* (FOR8) conidial formulations on the area under progress curve for the number of conidia of*Monilinia* spp. (AUncMOPC) and the number of *E. nigrum* or *P. frequentans* conidia (AUncPC), and the area under progress curve for the number of colony-forming units (CFU) of *E. nigrum* or *P. frequentans* (AUcfuPC) for the surfaces of "Autumn Free" nectarines from a commercial orchard in Sudanell, Spain in 2004 and 2005.

Year	Formulation	AUncMOPC	AUncPC	AUcfuPC
	FOR7	2,663,570 a	4.29×10^6 a	33,753 a
2004	FOR8	621,875 b	2.54×10^7 b	644,628 c
	cyproconazole	2,559,080 a	-	-
	control	3,121,590 a	-	-
	FOR7	356,349 b	8.27×10^6 a	43,995 a
2005	FOR8	59,868 c	1.92×10^7 b	138,062 b
	tebuconazole	308,256 b	-	-
	control	419,585 b	-	-
	MSE	1.56×10^{14}	1.71×10^{14}	1.31×10^{10}

*Data are the mean of four replicates. Data were analyzed by analysis of variance. Means followed by the same letter in each column were not significantly different according to the Student–Newman–Keuls multiple range test (p=0.05). MSE, mean squared error.

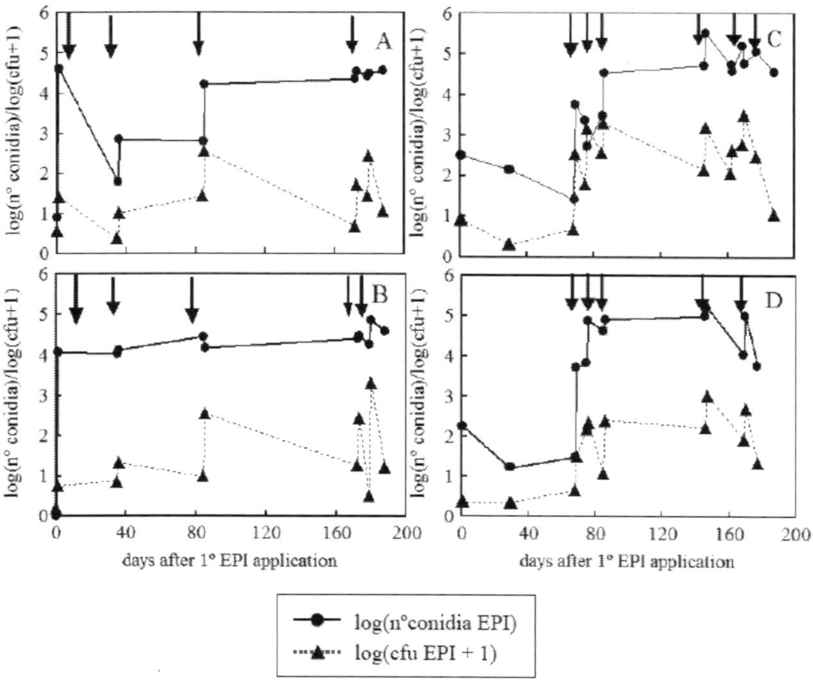

Figure 1. Population dynamics of *Epicoccum nigrum* (EPI) after five or six applications of the FOR7 *E. nigrum* conidial formulation. The data are displayed as the logarithm of the number of conidia and the number of colony-forming units (CFU + 1) that were recovered from flowers or fruit in (A) 2004 and (B) 2005 in a commercial orchard in Alfarrás, Spain, and in (C) 2004 and (D) 2005 in a commercial orchard in Sudanell, Spain. The number of *E. nigrum* conidia was counted in a hemocytometer under a light microscope (×100). The numbers of CFUs of *E. nigrum* per square centimeter of surface of the Petri dish that contained potato dextrose agar that was supplemented with 0.5 g L^{-1}streptomycin were counted with the naked eye. Data are the mean of four replicates, with 10 flowers or five fruits per replicate. The vertical arrows show the times of the EPI applications.

A principal component analysis (PCA) was undertaken to extract the linear composites of the number of *Monilinia* spp., *E. nigrum* and *P. frequentans* conidia on the nectarine and peach surfaces, the number of applications, the year of application, the formulation type, the tm, and the RH (Figure 3). According to the results of the PCA, we found that only the first three components had eigenvalues greater than one, and when combined, components 1, 2, and 3 accounted for 77.6% of the total variance, 41.3%, 23.5%, and 12.9%, respectively. The variance in the number of *Monilinia* spp. on fruit surface is mainly dependent on the number of *E. nigrum* and *P. frequentans*, the formulation type, and the RH (Figure 4). All were on the negative side of the first principal component axis, while the number of applications, the year of application, and the tm were on the positive side (Figure 4). The AUncMOPC and climatic conditions were on the negative side of the second principal component (Figure 4). On other hand, the AUncMOPC, the AUcfuPC, the RH, and the formulation type were closely grouped on the positive side of the third principal component axis (Figure 4). Shaw *et al.* [44] reported that the severity of a fungal disease in fruit is a function of temperature and wetness duration under controlled conditions. De Cal *et al.* [45] studied the population dynamics of *E. nigrum* on peaches and nectarines in order to find a method for promoting its colonization on fruit surfaces and improve its biocontrol efficacy against brown rot. They reported that the number of pit hardening and preharvest applications of the *E. nigrum* conidial formulations and RH accounted for 96% of the variability in the viability of the *E. nigrum* conidial population on the peach surfaces. In another study, Guijarro *et al.* [46] studied the population dynamics of *P. frequentans* on peach flower and fruit surfaces after different field treatments of six different *P. frequentans* conidia formulations. They found that size the *P. frequentans* conidial population on the surfaces was dependent upon application of the formulation to the blooming flowers and pre-harvested fruit, the RH, and the tm.

Preharvest application and optimization of BCAs require considerable understanding of the crop system, the pathogen's epidemiology, and the biology, ecology, and population dynamics of the antagonists and their interactions [47]. The results of our studies suggest that the size of *P. frequentans* population after an application of a *P. frequentans* conidial formulation during the crop season is larger than that of *E. nigrum* following the application of an *E. nigrum* conidial formulation. Moreover, applications of a *P. frequentans* conidial formulation during the crop season also caused a greater reduction in the number of *Monilinia* spp. conidia on the fruit surface than that found after applications of an *E. nigrum* formulation during the growing season.

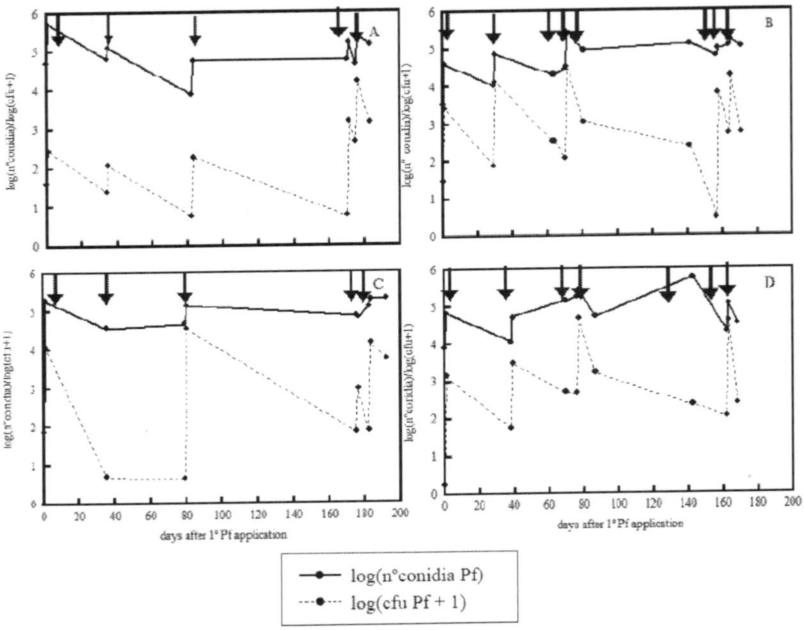

Figure 2. Population dynamics of *Penicillium frequentans* (Pf) after five or eight applications the FOR8 *P. frequentans* conidial formulation. The data are displayed as the logarithm of the number of conidia and the number of colony-forming units (CFU + 1) that were recovered from flowers or fruit in 2004 (A) and 2005 (B) in a commercial orchard in Alfarrás, Spain, and in 2004 (C) and 2005 (D) from a commercial orchard in Sudanell, Spain. The numbers of *P. frequentans* conidia were counted in a hemocytometer under a light microscope (×100). The number of CFUs of *P. frequentans*per square centimeter of surface of the Petri dish containing potato dextrose agar that was supplemented with 0.5 g L^{-1} streptomycin was counted with the naked eye. Data are the mean of four replicates, with 10 flowers or five fruits per replicate. The vertical arrows show the times of the Pf applications.

3. EXPERIMENTAL SECTION

3.1. Cultures and Formulations

Conidia of *E. nigrum* (ATCC No 96794) and *P. frequentans* (ATCC No 66108) were produced in a solid fermentation system [48,49]. For production of the FOR7 *E. nigrum* conidial formulation, fresh *E. nigrum* conidia were suspended after centrifugation in a 1% KCl + 1.25% sodium alginate solution in 250 mL centrifuge tubes. The tubes were shaken on a vortex mixer (Reax Top, Heildolph, Rose Scientific Ltd. Alberta, Canada) for 10 seconds and kept for 10 min. at room temperature (20–25°C). For production of the FOR8 *P. frequentans* conidial formulation, fresh *P. frequentans* conidia were suspended in 2% methyl cellulose +10% glucose solution in 250 mL centrifuge tubes, which were then processed in the identical manner as described above. Each conidial suspension was filtered through 2µm filter paper in a Büchner funnel, and the filtrate was then mixed with silica powder at a ratio that was necessary to obtain granulation [26,27]. The granules were dried in a fluid bed dryer (FBD model 350s, Burkard Manufacturing Co Ltd., Hertfordshire, UK) at its highest air flow rate at 40°C [25,50]. The moisture content of each final conidial formulation was measured using a humidity analyzer (BOECKEL, GmbH & Co, Hamburg, Germany). Germination of the dried conidia was assessed using a previously described bioassay [25,50]. Conidia were dried to reduce their moisture content to between 5% and 10% and their germinability was as high as 80% after drying [26,50].

3.2. Treatments and Experimental Design

Four field surveys were carried out in two commercial peach (*Prunus persica* "Red September") and nectarine (*Prunus persica* var. *nectarina* (Aiton) Maxim "Autumn Free") orchards which were located in Alfarrás (UTM: 298197, 4634045), and Sudanell (UTM: 297230, 4603615), Lleida, Cataluña, Spain over two growing seasons in 2004 and 2005. The size of the orchards ranged from 3–5 hectares. The planting distances between rows and trees ranged from 2–4 metres in the Alfarras orchard, and 4–5 metres in the Sudanell orchard. The experimental design of each experiment was a randomized complete block design in which four replications (four blocks) were used, and three trees comprised the sample unit for each treatment and replication. Three barrier trees were used to separate the blocks and the treatments in order to avoid inter plot interference and protect the untreated trees from the fungicidal sprays. Daily measurements of mean tm, rainfall (mm), and RH were collected by automated weather monitors, which were located 0.5–5 km from each orchard. The average daily tm, rainfall, and RH were calculated for each orchard and for each treatment period. The treatment periods were (a) day of first biocontrol treatment to one day before the second treatment; (b) day of second biocontrol treatment to one day before the third treatment; (c) day of third biocontrol treatment to one day before the fourth treatment, and (d) day of fourth biocontrol treatment to one day before the fifth treatment. The trees were treated 5–8 times with the FOR7 and FOR8 conidial formulations in each experiment during the growing season: once at the pink blossom stage, once at the shuck-split stage, 1–3 times at the pit hardening stage, and 1–3 times at preharvest. All treatments were applied using a backpack sprayer (operating pressure 10 bar, hollow cone nozzle 1 mm) in the morning. Standard crop protection practices were applied in the two commercial orchards during the surveys. No specific fungicide against *Monilinia* spp. was applied during the surveys, except when trees in the same orchard were treated with either an experimental conidial formulation or a fungicide. When applied, the specific fungicides were: Caddy Pepite 10 (cyproconazole 10% WG; Bayer Cropscience, S.L. Valencia, Spain) in 2004 (ALF04 and SUD04); and Folicur 25RW (tebuconazole 25% WG; Bayer Hispania Industrial, S.A., Barcelona, Spain) in 2005 (ALF05 and SUD05). All treatments were carefully applied in non-windy conditions.

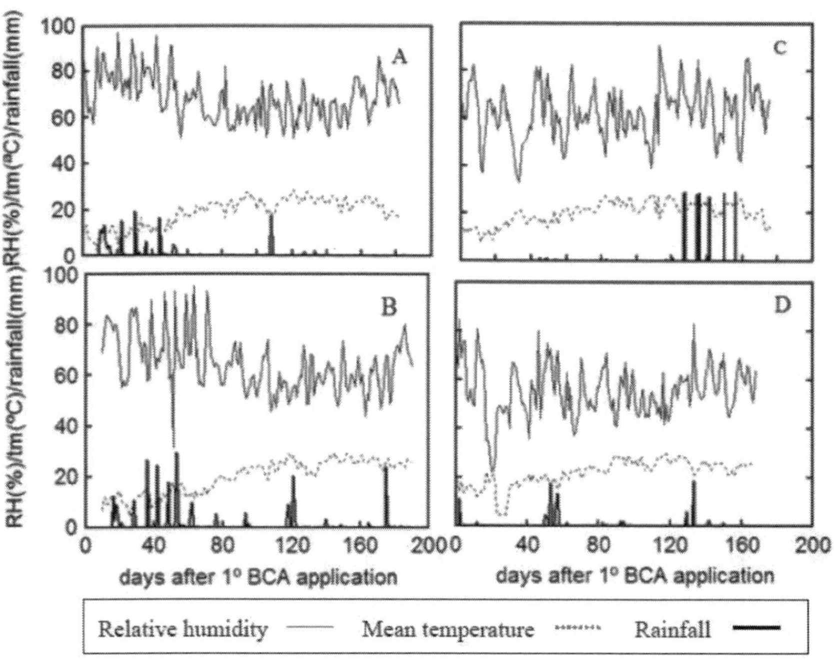

Figure 3. Daily rainfall (mm), mean temperature (tm ºC) and relative humidity (% RH) during (A) 2004 and (B) 2005 in the commercial orchard in Alfarrás, Spain, and (C) 2004 and (D) 2005 in the commercial orchard in Sudanell, Spain.

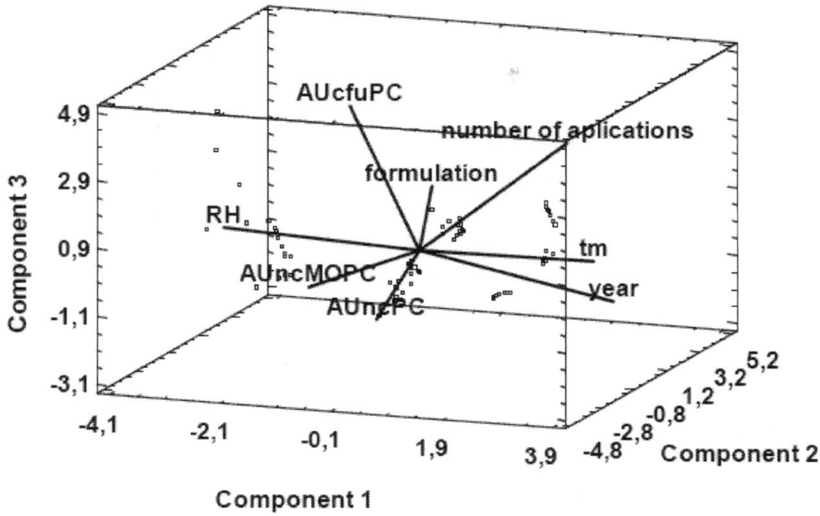

Figure 4. Principal component analysis to the number of *Monilinia* spp. (AUncMOPC), the number of *E. nigrum* and *P. frequentans* conidia and cfu (AUncPC and AUcfuPC) on nectarine and peach surfaces, the number of applications, the year of application, the formulation type, the mean temperature (tm), and the relative humidity (RH). The first three components accounted for 41.3%, 23.5%, and 12.9% of the total variance, respectively.

3.3. Population Dynamics of E. nigrum and P. frequentans

In order to investigate the population dynamics of *E. nigrum* and *P. frequentans* following the application of each conidial formulation, 10 flowers or five fruits per sample unit (a sample unit corresponded to three trees and four replications for each treatment) were sampled in each orchard. The fruit and flower samples were taken at random different heights and locations in the canopy of each tree in a manner to ensure that the entire canopy of each tree was sampled. The sampling was undertaken 11 times in 2004 and 12 times in 2005. In the laboratory, each sample unit was suspended in containers that contained sterile distilled water (SDW) and shaken for 30 min. at 150 rev min^{-1} in a rotary shaker. Each suspension was then decanted to centrifuge tubes, which were then centrifuged at 4°C for 10 min. at 14,040*g*. The recovered pellet was resuspended in 5 mL SDW (concentrate). The population size of *E. nigrum* and *P. frequentans* in each suspension was estimated from (i) the number of conidia and (ii) the number of CFUs of *E. nigrum* and *P. frequentans* per flower or fruit. The numbers of *E. nigrum* and *P. frequentans* conidia were counted in a hemocytometer under a light microscope (×100). The number of CFUs of *E. nigrum* and *P. frequentans* per Petri dish that contained PDA (Difco, Detroit, MI, USA) which was supplemented with 0.5 g L^{-1} streptomycin was counted with the naked eye. Aliquots (100 µL) of 10- and 100-fold dilutions of each concentrate and each concentrate were spread onto the Petri dishes that contained the amended PDA. Three replicate dishes were used for each replicate, dilution, and concentrate. The Petri dishes were maintained at 20–25°C for 5–7 days in the dark before counting the number of CFUs [51]. The control for this assay was flowers and fruit from the Sudanell orchard which did not receive any treatment in 2005 (SUD05) and flowers and fruit from the Alfarrás orchard in 2005 (ALF05). Morphological characterization of *E. nigrum* and *P. frequentans* conidia in the CFUs was used to confirm their identity.

3.4. Biocontrol Efficacy of E. nigrum and P. frequentans Conidial Formulations

In order to evaluate the effect of *E. nigrum* and *P. frequentans* conidial formulations on the number of *Monilinia* spp. conidia on peach surfaces, we used the same flower and fruit samples that were used to estimate the number of *P. frequentans* and *E. nigrum* on the flower or fruit samples.

3.5. Data Analysis

Progress curves for the conidial numbers of *E. nigrum, P. frequentans,* and *Monilinia* spp., and for CFUs of *E. nigrum* and *P. frequentans* for each treatment at the different sampling dates were generated for each orchard and year. The area under each progress curve for the number of *E. nigrum* and *P. frequentans* conidia (AUncPC) or *Monilinia* spp. conidia (AUnc MOPC), and the number of CFUs of *E. nigrum and P. frequentans* (AUcfuPC) for each orchard and year were calculated by trapezoidal integration on an Excel spreadsheet using the following formulae that are based on those used by Campbell and Madden [52]:

$$AUncPC = \Sigma_{i=1}^{n-1}[(t_{i+1} - t_i)(NC_i + NC_{i+1})/2] \quad (1)$$

$$AUncMONPC = \Sigma_{i=1}^{n-1}[(t_{i+1} - t_i)(NCMO_i + NCMO_{i+1})/2] \quad (2)$$

$$AUcfuPC = \Sigma_{i=1}^{n-1}[(t_{i+1} - t_i)(CFU_i + CFU_{i+1})/2] \quad (3)$$

where "*t*" is "days from full flowering to harvest", "NC" or "NCMO" are "the number of conidia at each sampling date for *E. nigrum* and *P. frequentans* or *Monilinia* spp., respectively", "CFU" is "the number of CFUs at each sampling date for *E. nigrum* and *P. frequentans*", and "*n*" is "the number of sampling dates".

Data of the AUncMOPC, and AUncPC and the AUcfuPC for each formulation and for each orchard were analyzed by a factorial analysis of variance [53]. When the F-test showed a significant difference at $p \leq 0.05$, the treatment means were compared by Student–Newman–Keuls multiple range test (p = 0.05) [53].

The number of *Monilinia* spp., *E. nigrum* and *P. frequentans* conidia on nectarine and peach surfaces and climatic conditions from all experiments were analyzed by a principal component analysis [54] and only those components whose eigenvalues were greater or equal to 1.0 were extracted. The analysis was done with Statgraphics Centurion XVI v. 16.1.03 (StatPoint, Inc., Herndon, VA, USA).

4. CONCLUSIONS

Biological control of brown rot in stone fruit caused by *Monilinia* spp. is very dependent on BCA density on the flower or fruit surface and the RH. One way to improve the performance of a fungal BCA is to develop an efficacious conidial formulation. The size of *P. frequentans* population after an application of a *P. frequentans* conidial formulation during the crop season is bigger than that of *E. nigrum* following the application of an *E. nigrum* conidial formulation. Moreover, applications of a *P. frequentans* conidial formulation during the crop season also caused a greater reduction in the number of*Monilinia* spp. conidia on the fruit surface than that found after applications of an *E. nigrum* formulation during the growing season.

Finally, the potency of the *E. nigrum* conidial formulation is the same as that of fungicide.

ACKNOWLEDGMENTS

This work was carried out with financial support from AGL2002-4396-CO2 (Plan Nacional de I+D+I, Ministerio de Educación y Ciencia, Spain), RTA2005-0077-CO2, and from RM2010-00013. The authors wish to thank Y. Herranz and M.T. Morales Clemente for their technical support. The authors also wish to thank Arieh Bomzon, Consulwrite (www. consulwrite. com) for his editorial assistance in preparing this review.

REFERENCES

1. Alabouvette, C.; Olivain, C.; Steinberg, C. Biological control of plant diseases: The European situation. *Eur. J. Plant Pathol.* **2006**, *114*, 329–341.

2. Nicot, P.C.; Decognet, V.; Fruit, L.; Bardin, M.; Trottin, Y. Combined effect of microclimate and dose of application on the efficacy of biocontrol agents for the protection of pruning wounds on tomatoes against *Botrytis cinerea*. *Bulletin IOBC/SROP* **2002**, *25*, 73–76.

3. Byrde, R.J.; Willetts, H.J. *The Brown Rot Fungi of Fruit. Their Biology and Control.*, 1st ed; Pergamon Press: Oxford, UK, 1977; p. 171.

4. Ahmadi, H.; Biasi, W.V.; Mitcham, E.J. Control of brown rot decay of nectarines with 15% carbon dioxide atmosphere.*J. Am. Soc. Hort. Sci.* **1999**, *124*, 708–712.

5. Sommer, N.F. Role of controlled environments in suppression of post-harvest diseases. *Can. J. Plant Pathol.* **1985**, *7*, 331–339.

6. Osorio, J.M.; Adaskaveg, J.E.; Ogawa, J.W. Comparative efficacy and systemic activity of iprodione and the experimental anilide E-0858 for control of brown rot on peach fruit. *Plant Dis.* **1993**, *77*, 1140–1143.

7. Hong, C.X.; Michailides, T.J.; Holtz, B.A. Effects of wounding, inoculum density, and biological control agents on postharvest brown rot of stone fruit. *Plant Dis.* **1998**, *82*, 1210–1216.

8. Ogawa, J.M.; Zehr, E.I.; Biggs, A.R. Brown rot. In *Compendium of Stone Fruit Diseases*; Ogawa, J.M., Zehr, E.I., Bird, G.W., Ritchie, D.F., Uriu, K., Uyemoto, J.K., Eds.; APS Press: St. Paul, MN, USA, 1995; pp. 7–10.

9. Hong, C.X.; Holtz, B.A.; Morgan, D.P.; Michailides, T.J. Significance of thinned fruit as a source of the secondary inoculum of *Monilinia fructicola* in California nectarine orchards. *Plant Dis.* **1997**, *81*, 519–524.

10. Villarino, M.; Melgarejo, P.; Usall, J.; Segarra, J.; Lamarca, N.; De Cal, A. Secondary inoculum dynamics of *Monilinia*spp. and relationship to the incidence of postharvest brown rot in peaches and the weather conditions during the growing season. *Eur. J. Plant Pathol.* **2012**.

11. Xu, X.-M.; Bertone, C.; Berrie, A. Effects of wounding, fruit age and wetness duration on the development of cherry brown rot in the UK. *Plant Pathol.* **2007**, *56*, 114–119.

12. Gell, I.; De Cal, A.; Torres, R.; Usall, J.; Melgarejo, P. Relationship between the incidence of latent infections caused by*Monilinia* spp. and the incidence of brown rot of peach fruit: Factors affecting latent infection. *Eur. J. Plant Pathol.* **2008**,*121*, 487–498.

13. Villarino, M.; Sandin-España, P.; Melgarejo, P.; De Cal, A. High chlorogenic and neochlorogenic acid levels in immature peach reduce *Monilinia laxa* infection by interfering with fungal melanin biosynthesis. *J. Agr. Food Chem.***2011**, *59*, 3205–3213.

14. Penrose, L.J.; Davis, K.C.; Koffmann, W. The distribution of benomyl-tolerant *Slerotinia fructicola* (Wint.) Rehm in stone fruit orchards in New South Wales and comparative studies with susceptible isolates. *Aust. J. Agric. Res.* **1979**,*30*, 307–319.

15. Penrose, L.J. Prolonged field persistence of resistance to benomyl in *Monilinia fructicola*. *Crop Prot.* **1990**, *9*, 190–192.

16. Elmer, P.A.G.; Gaunt, R.E. Effect of frequency of dicarboximide applications on resistant populations of *Monilinia fructicola* and brown rot in New Zealand orchards. *Crop Prot.* **1993**, *12*, 83–88.

17. Falconi, C.J.; Mendgen, K. Epiphytic fungi on apple leaves and their value for control of the postharvest pathogens*Botrytis cinerea*, *Monilinia fructigena* and *Penicillium expansum*. *J. Plant Dis. Prot.* **1994**, *101*, 38–47.

18. Sanoamuang, N.; Gaunt, R.E. Persistence and fitness of carbendazim- and dicarboximide-resistant isolates of *Monilinia fructicola* (Wint.) Honey in flowers, shoots and fruit of stone fruit. *Plant Pathol.* **1995**, *44*, 448–457.

19. Adaskaveg, J.E.; Förster, H.; Gubler, W.D.; Teviotdale, B.L.; Thompson, D.F. Reduced risk fungicides: A new way of managing brown rot and other fungal diseases of stone fruit crops. *Cal. Agric.* **2004**, *59*, 109–114.

20. Malavolta, C.; Cross, J.V.; Cravedi, P.; Jörg, E. Guidelines for Integrated Production of Stone Fruits. In *Technical Guideline III, IOBC wprs Bulletin*; International Organization for Biological and integrated Control of Noxious Animals and Plants, West Palaearctic Regional Section: Dijon, France, 2003; Volume 26.

21. Melgarejo, P.; Carrillo, R.; Sagasta, E.M. Mycoflora of peach twigs and flowers and its possible significance in biological control of *Monilinia laxa*. *Trans. Br. Mycol. Soc.* **1985**, *85*, 313–317.

22. Melgarejo, P.; Carrillo, R.; Sagasta, E.M. Potential for biological control of *Monilinia laxa* in peach twigs. *Crop Prot.***1986**, *5*, 422–426.

23. De Cal, A.; Sagasta, E.M.; Melgarejo, P. Biological control of peach twig blight (*Monilinia laxa*) with *Penicillium frequentans*. *Plant Pathol.* **1990**, *39*, 612–618.

24. Madrigal, C.; Pascual, S.; Melgarejo, P. Biological control of peach twig blight (*Monilinia laxa*) with *Epicoccum nigrum*.*Plant Pathol.* **1994**, *43*, 554–561.

25. Guijarro, B.; Larena, I.; Melgarejo, P.; De Cal, A. Effect of drying on conidial viability of *Penicillium frequentans*, a biological control agent against peach brown rot disease caused by *Monilinia* spp. *Biocontr. Sci. Technol.* **2006**, *16*, 257–269.

26. Guijarro, B.; Melgarejo, P.; De Cal, A. Effect of stabilizers on the shelf-life of *Penicillium frequentans* conidia and their efficacy as a biological

agent against peach brown rot. *Int. J. Food Microbiol.* **2007**, *113*, 117–124.

27. Larena, I.; De Cal, A.; Melgarejo, P. Effects of stabilizers on shelf-life of Epicoccum nigrum formulations and their relationship with biocontrol of postharvest brown rot by *Monilinia* of peaches. *J. Appl. Microbiol.* **2007**, *102*, 570–582.

28. De Cal, A.; Sagasta, E.M.; Melgarejo, P. Antifungal substances produced by *Penicillium frequentans* and their relationship to the biocontrol of *Monilinia laxa*. *Phytopathol.* **1988**, *78*, 888–893.

29. De Cal, A.; Melgarejo, P. Interactions of and mycoflora of peach twigs. *Mycol. Res.* **1992**, *96*, 1105–1113.

30. Janisiewicz, W.J.; Jeffers, S.N. Efficacy of commercial formulation of two biofungicides for control of blue mold and gray mold of apples in cold storage. *Crop Prot.* **1997**, *16*, 629–633.

31. Guijarro, B.; Melgarejo, P.; De Cal, A. Effects of different biological formulations of *Penicillium frequentans* on brown rot of peaches. *Biol. Control* **2007**, *42*, 86–96.

32. Burges, H.D.; Jones, K.A. Trends in formulation of microorganisms and future research requirements. In *Formulation of Microbial Biopesticides*, 1st; Burges, H.D., Ed.; Kluwer Academic Publishers: Dordrecht, The Netherlands, 1998; pp. 311–332.

33. Collins, D.P.; Jacobsen, B.J.; Maxwell, B. Spatial and temporal population dynamics of a phyllosphere colonizing*Bacillus subtilis* biological control agent of sugar beet Cercospora leaf spot. *Biol. Control* **2003**, *26*, 224–232.

34. Jones, E.E.; Mead, A.; Whipps, J.M. Effect of inoculum type and timing of application of Coniothyrium minitans on Sclerotinia sclerotiorum: Control of sclerotinia disease in glasshouse lettuce. *Plant Pathol.* **2004**, *53*, 611–620.

35. Jones, E.E.; Clarkson, J.P.; Mead, A.; Whipps, J.M. Effect of inoculum type and timing of application of *Coniothyrium minitans* on *Sclerotinia sclerotiorum*: Influence on apothecial production. *Plant Pathol.* **2004**, *53*, 621–628.

36. Gell, I.; De Cal, A.; Torres, R.; Usall, J.; Melgarejo, P. Conidial density of *Monilinia* app. on peach fruit surfaces in relation to the incidences of latent infections and brown rot. *Eur. J. Plant Pathol.* **2009**, *123*, 415–424.

37. Larena, I.; Torres, R.; De Cal, A.; Liñán, M.; Melgarejo, P.; Domenichini, P.; Bellini, A.; Mandrin, J.F.; Lichou, J.; Ochoa de Eribe, X.; Usall, J. Biological control of post-harvest brown rot (*Monilinia* spp.) of peaches by field applications of*Epicoccum nigrum*. *Biol. Control* **2005**, *32*, 305–310.

38. Kurtzman, C.; Droby, S. *Metschnikowia fructicola*, a new ascosporic yeast with potential for biocontrol of postharvest fruit rots. *Syst. Appl. Microbiol.* **2001**, *24*, 395–399.

39. Smilanick, J.L.; Dennis-Arue, R.; Bosch, J.R.; Gonzales, A.R.; Henson, D.; Janisiewicz, W.J. Control postharvest rot of nectarines and peaches by *Pseudomonas* species. *Crop Prot.* **1993**, *12*, 513–520.

40. Kinkel, L.L. Microbial population dynamics on leaves. *Annu. Rev. Phytopathol.* **1997**, *35*, 327–347.

41. Cartwright, K.D.; Benson, D.M. Optimization of biological control of *Rhizoctonia* stem rot on poinsettia by *Paecilomyces lilacinus* and *Pseudomonas cepacia*. *Plant Dis.* **1995**, *79*, 301–308.

42. Domsch, K.H.; Gams, W.; Anderson, T.H. *Compendium of Soil Fungi*, 1st ed; Academic Press: New York, NY, USA, 1980; p. 1264.

43. Teixidó, N.; Viñas, I.; Usall, J.; Magan, N. Control of blue mold of apples by preharvest application of *Candida sake*grown in media with different water activity. *Phytopathol.* **1998**, *88*, 960–964.

44. Shaw, D.A.; Adaskaveg, J.E.; Ogawa, J.M. Influence of wetness period and temperature on infection and development of shot-hole disease of almond caused by *Wilsonomyces carpophilus*. *Phytopathol.* **1990**, *80*, 749–756.

45. De Cal, A.; Larena, I.; Liñán, M.; Torres, R.; Lamarca, N.; Usall, J.; Domenichini, P.; Bellini, A.; Eribe, X.O.; Melgarejo, P. Population dynamics of *Epicoccum nigrum*, a biocontrol agent against brown rot in stone fruit. *J. Appl. Microbiol.***2009**, *106*, 592–605.

46. Guijarro, B.; Melgarejo, P.; Torres, R.; Lamarca, N.; Usall, J.; De Cal, A. *Penicillium frequentans* population dynamic on peach surfaces after its application against brown rot in orchards. *J. Appl. Microbiol.* **2008**, *104*, 659–671.

47. Andrews, J.H. Biological control in the phyllosphere. *Annu. Rev. Phytopathol.* **1992**, *30*, 603–635.

48. De Cal, A.; Larena, I.; Guijarro, B.; Melgarejo, P. Mass production of conidia of *Penicillium frequentans*, a biocontrol agent against brown rot of stone fruits. *Biocontrol Sc. Technol.* **2002**, *12*, 715–725.

49. Larena, I.; De Cal, A.; Melgarejo, P. Solid substrate production of *Epicoccum nigrum* conidia for biological control of brown rot on stone fruits. *Int. J. Food Microbiol.* **2004**, *94*, 161–167.

50. Larena, I.; De Cal, A.; Liñán, M.; Melgarejo, P. Drying of *Epicoccum nigrum* conidia for obtaining a shelf-stable biological product against brown rot disease. *J. Appl. Microbiol.* **2003**, *94*, 508–514.

51. Lacey, J.; Hill, S.T.; Edwards, M.A. Microorganisms in stored grains, their enumeration and significance. *Trop. Stored Prod. Inf.* **1980**, *39*, 19–33.

52. Campbell, C.L.; Madden, L.V. *Introduction to Plant Disease Epidemiology*, 1st ed; Wiley-Interscience: New York, NY, USA, 1990; p. 523.

53. Snedecor, G.W.; Cochran, W.G. *Statistical Methods*, 7th ed; Iowa State University Press: Ames, IA, USA, 1980.

54. Madden, L.; Pennypacker, S.P. Principal Component analysis of tomato early blight epidemics. *Phytopathol. Z. J. Phytopathol.* **1979**, *95*, 364–369

Chapter 3

Novel Techniques for Gene Delivery into Plants and Its Applications for Disease Resistance in Crops

Ahmad Hama Ameen Rashid, Djshwar Dhahir Lateef

Field Crop Departments, College of Agriculture, University of Sulaimani, Sulaimaniyah, Iraq

ABSTRACT

From the early past to the present, biotechnologies have produced the ability to genetically transform a wide variety of plant species. The plant transformation technologies have changed the face of agriculture and plant biology. Plant genetic transformation is one of the key technologies for crop improvement in addition to emerging approach for producing recombinant proteins in plants. Both plastid genomes and plant nuclear can be genetically modified. Until now, essential functional differences between the prokaryotic-like genome of the plastid and the eukaryotic genome of the plant cell nucleus will have an impact on characteristics of transgenic organism. Thus, the main goals are to generate transgenic plants with the traits of interest as well as minimizing the amount of transgenic DNA in plants while maximizing stability of gene expression and trait performance. In this review, two broad groups of gene delivery methods will be discussed namely, (bilogical and physical methods) and subsequently there applications for improving disease resistance will be discussed.

KEYWORDS

T-DNA, Particle Bombardment, MSNs, Silicon Carbide-Mediated Transformation, Electroporation

1. INTRODUCTION

Fundamental element for crop improvement and development of new cultivars with desirable characters is the ability to modify the genetic composition of a plant. To develop new lines by sexual hybridization, plant breeders have employed the naturally happening genetic variability in existing germplasm [1].

To generate genetic variability, the development of somatic cell hybridization and protoplast culture is one of the first patterns. Additionally, in vitro culture of plant cells in suboptimal conditions is established to induce genetic variations which term somaclonal variation [2]. To create genetic variation across species, the Agrobacterium tumefaciens-mediated integration of foreign DNA into a cell's nuclear genome has become the method of choice in which the introduced gene is inherited following Mendelian genetics, regardless of sexual compatibility or genetic proximity [3].

The first step for genetic transformation is the DNA delivery to cells or tissues, followed by culture and selection to permit only those tissues and cells having a marker gene (for instance, herbicide resistance gene or antibiotic) to proliferate and survive further. Then, plants are regenerated from the surviving cells, embedded and rooted in the soil. Such primary transformants after that are used for molecular analyses for determination of copy number of the transgenes of interest [4] [5] (For more details, see Figure 1).

Approaches for delivering DNA into plant cells and gene transformation can be divided into two major categories: direct and indirect DNA deliveries. Direct method does not employ bacterial cells as mediators, as an alternative, it uses a chemical alteration or physical force such as electric discharge or pressure to deliver the vector DNA into a host cell [5]. Whereas, in indirect approach, genes of interest are introduced into the target cell through bacteria, for instance, Agrobacterium tumefaciens or Agrobacterium rhizogenes [1]. For introducing genes into plants, microprojectile bombardment and Agrobacterium-mediated gene transfer direct are the most widely exploited methods because of their ability to transform intact, re-generable tissues and organs [6] -[8]. In this essay, two broad groups of gene delivery methods will be discussed namely (bilogical and physical methods), and subsequently there applications for improving disease resistance will be discussed.

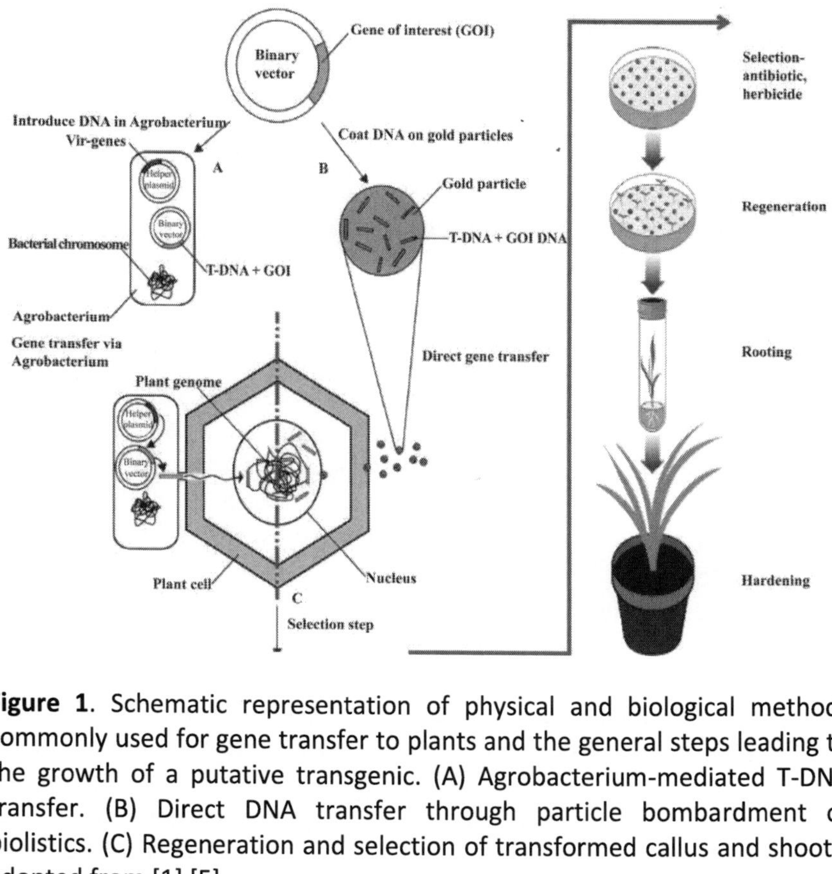

Figure 1. Schematic representation of physical and biological methods commonly used for gene transfer to plants and the general steps leading to the growth of a putative transgenic. (A) Agrobacterium-mediated T-DNA transfer. (B) Direct DNA transfer through particle bombardment or biolistics. (C) Regeneration and selection of transformed callus and shoots, adopted from [1] [5].

2. BIOLOGICAL METHOD

2.1 Agrobacterium-Mediated Gene Transfer

Agrobacterium tumefaciens is a Gram-negative soil phytopathogenic bacterium which causes crown gall disease in plants and it can be grown in vitro in simple culture media without any added plant growth hormones and the bacterium [9]. This disease is established by neoplastic growth caused by the integration of the transferred DNA (T-DNA) obtained from specific DNA fragments (Ti) plasmid into the plant nuclear genome [10]. This feature is widely used in plant biotechnology, and Agrobacterium is the most important tool employed to produce transgenic plants [5]. In the past three decades, the biology of Agrobacterium and its relations with host plant have been the subject of many studies, for recent review see [11].

The understanding of the T-DNA integration pathways have considerably enriched by releasing new studies and in the same time, it exposed many host factors that participate in these events [7] [11]. T-DNA integration for Agrobacterium includes two main steps: firstly, the T-strand is converted to a double-stranded form; and secondly, the host cell DNA repair machinery mediates the double-stranded T-DNA integration into double strand breaks in the host genome [11].

Briefly, the following steps of host genetic transformation facilitated by Agrobacterium are: The induction of Agrobacterium's virulence machinery results in expression and activation of the virulence genes (vir genes), some of these genes are important in the transfer of T-DNA from the bacterium to host cell, whereas others helps in targeting T-DNA to the nucleus and most likely to the precise integration site in the host cell for T-DNA [5] [12]. This first step mobilises a single stranded DNA segment from the (Ti) plasmid which usually contains one T-DNA region. However, in some cases, a Ti plasmid may comprise multiple T-DNA regions [13]. This section of transferred DNA (T-DNA), bounded by two (25) bp direct repeat sequences defined as right and left borders (RB and LB) [14], it is termed the T-strand, and the DNA present in between the border sequences is transferred to the receiver plant nucleus [11]. In addition, VirD2, linked with VirD1, forms a nuclease able to remove the T-strand by a strand-replacement mechanism, at the completion of which VirD2 remains covalently linked to the 5-end (RB) of the T-strand [14]. For more details, see Figure 2.

This method has been found to be challenging due to the low copy number and large size of Ti plasmids, leading to difficulties in plasmid manipulation and isolation, and it limited in the range of plant species that can be transformed because not all tissues or species are susceptible to Agrobacterium [15] [16]. To support these limitations, it has been reported by the group of researchers in Japan and Australia Nakano, Suzuki [17], large DNA molecules about 100 kb integrated by this indirect method and tend to be rearranged with deletions, duplications and insertions not only in the transgenic rice lines but also in itself. Therefore, this method may not be the ideal for transformation large DNA fragments.

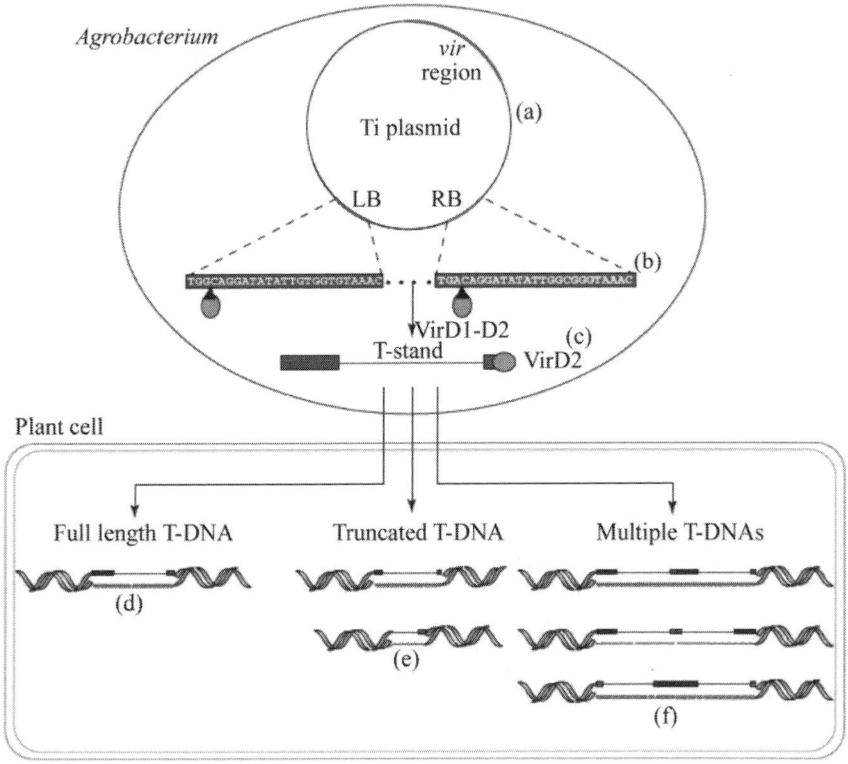

Figure 2. Structure representation of the Agrobacterium Ti-plasmid and integrated T-DNA molecules. (a) The T-DNA section includes left border (LB), right border (RB) and the virulence (vir) region. (b) The borders of T-DNA are 25-bp repeats and works as targets for the VirD1 - VirD2 endonuclease complex (c) As single-stranded DNA molecule, the (T-strand) is released and T-DNA typically take part into the host genome (d) In various orientationsas a single full-length or (e) truncated molecule in addition to (f) multiple molecules joined to each other [4].

Beside some drawbacks for Agrobacterium transformation, this method still works in many labs and high transformation rate frequencies could be obtained. More recently, by group of researchers Chetty, Ceballos [18], four

Agrobacterium tumefaciens strains (GV3101, EHA105, AGL1, and MP90) for the genetic transformation of tomato (Solanum lycopersicum L.) cultivar was evaluated. The reasons they chose Micro-Tom for their experiment was due to the fact that it differs from typical tomato cultivars by having two recessive genes, which confer the dwarf phenotype and it is preferred as a model system for functional genomics, because it shares several unique features with Arabidopsis, such as small size and short life cycle [19]. Transformation rate frequencies for all strains were (65%, 40%, 35% and 15%) respectively. The analysis by comparative qPCR technique for duplex Taq-Man reactions showed that the strain MP90 was the most efficient in the transfer of single transgene insertions into the tomato genome regardless to their lowest transformation rate (see Table 1). Thus, their discovery could be used for functional genomics and biotechnological applications in tomato.

3. PHYSICAL METHODS

3.1. Microprojectile Bombardment

Particle bombardment is a theoretically simple powerful tool for biotechnologists allowing direct gene transfer to a broad range of cells and tissues which have been found difficult to transform by other technique, and is more valuable for improving species which have high level of heterozygosity such as potato and cassava [20].

The term biolistic was established to describe the nature of the delivery of the foreign DNA into living cells,and the development of the particle bombardments from concept to useful process was first invented by Sanford [21], in which gene carriers can penetrate virtually any type of tissue or at least the first few cell layers and to generate transgenic plants in species which were not amenable to Agrobacterium-mediated genetic transformation.

Principally, the particle bombardment device comprises of a mechanism to accelerate the particles to desired speeds and adjust their penetration into the receiver cells. Initially, the gun powder discharge device was used to accelerate inert metal microprojectiles coated with biologically active compounds [21]. Soon after, this device was replaced with inert gas helium (Biolistics1 PDS-1000/He) to offer the force for microprojection [22]. The rupture disk assembly is the major part of this most commonly used device which controls the helium pressure and helium gas released by a rupture disc and partial vacuum to propel a macrocarrier plastic sheet loaded with DNA coated tungsten or gold. This disk assembly contains a gas acceleration tube with a rupture disk placed at the bottom of the tube inside a retaining cap. Subsequently, the gas chamber is evacuated and the helium gas pressure is allowed to build up to the desired level to break the rupture disk. Finally, at the optimized velocity via metal screen, the optimized helium gas pressure forces the microprojectiles to deliver biologically active compounds and penetrate the target tissue to affect gene transfer [22] [23] (See Figure 3, for diagrammatic representation of this device). The main function of the microprojectiles is to create holes through which DNA passes into the cells. Therefore, a number of modifications have been made to the power source used to propel the microprojectiles to control microprojectile penetration into cells such as regulated nitrogen gas pressure, an air gun and compressed air [24] -[26]. A main difference in the air gun technique is that the DNA is suspended with the microprojectiles rather than being coated on them [25]. The creators of mentioned techniques assumed that the movement of DNA independent of microprojectiles which allows them to target small locations of tissues.

The critical factors to maximize transformation efficiency by this system are the delivery of optimal amount of DNA with minimal injury to the receiver plant tissue, size and density of microprojectiles, attachment of DNA to microprojectiles, helium pressure at which microprojectiles are propelled and distance between rupture disc are all important factors [4] [23]. Most of these parameters need to be optimized and differ with individual laboratories. By only changing the target diameter or adjusting the distance that the microparticles travel to the target tissue, transformation events may be affected [23].

One of the preferred methods for both transient gene expression studies and stable genetic transformation for improvement of crops is biolistic gene transfer [27]. For example, first transgenic sugarcane plants were generated by Bower and Birch [28], from embryonic callus, they easily recovered transgenic sugarcane plant with high DNA coated microprojectile by bombardment because the surface layer of sugarcane embryogenic callus obviously has a higher proportion of cells that are able to regenerate and proliferate that permit the selection of transformed plants [29]. Under the control of promoter Emu, transformation of a gene npt-II was delivered into sugarcane [30]. Subsequently, this transformation was confirmed by Southern hybridization.

Table 1. Percentage of transgenic plants based on a duplexed TaqMan qPCR assay with 1, 2 or 2 copies of nptII or uidA transgenes in their genomes, a total of 42 plants obtained with each Agrobacterium strain were used for the TaqMan assay [5].

Agrobacterium Strain	Transgene copy number (%)					
	nptII			*uidA*		
	1	2	>2	1	2	>2
AGL1	52	24	24	52	29	19
EHA105	67	21	12	62	25	13
GV3101	52	33	15	55	31	14
MP90	76	24	0	67	33	0

More recently for biolistic transformation of sugarcane Xiong, Jung [31], by using linear expression cassettes three different DNA coating procedures were compared, namely; spermidine, protamine sulfate and the Seashell DNAdelTM Gold Carrier with patented precipitation buffer from Seashell Technology. The free base (Spermidine) lead to 2.65-fold and three-fold higher transient GUS expression than Seashell DNAdelTM and protamine, respectively and they assumed that all three DNA coating protocols are similarly effective for stable genetic transformation of sugarcane (For more details, see Table 2 and Figure 4).

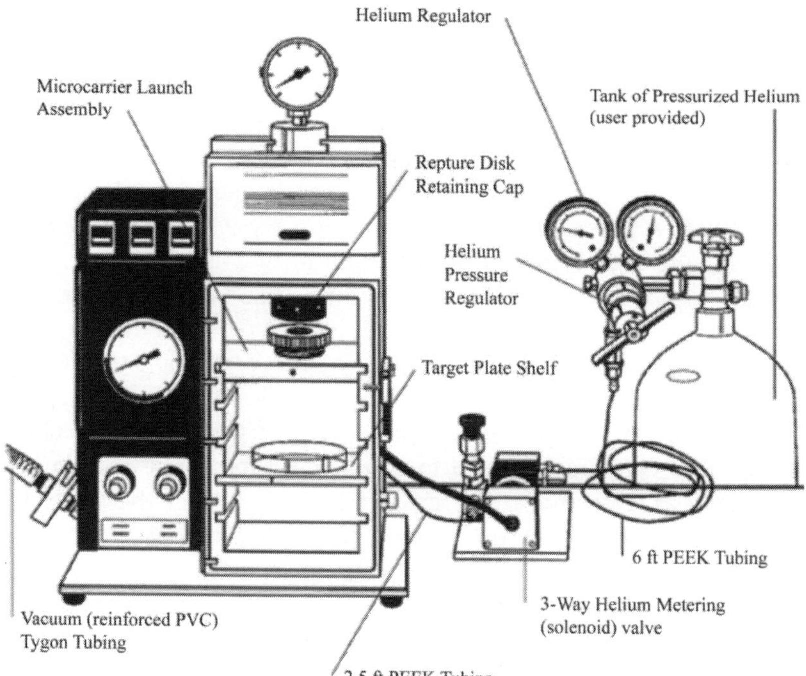

Figure 3. Diagrammatic representation of biolistic PDS-1000/He biolistic gun (courtesy biorad laboratories, USA) [1].

Recently, progress of a new active transgenesis method based on particle bombardment would be more effective than common particle bombardment and integration of multiple copies of DNA could be decreased [32].

Wu, Du [32] delivered DNA-enzyme complex into plant cells through the biolistic method by using 1 µm gold microparticles. The codelivery of the

DNA complex (DNA plasmid and protein) let to enhanced plant transformation efficiency. However; it required covalent modification of the protein so as to stay attached with the gold microparticle in the time of bombardment. Therefore, platform for codelivery the DNA complex to plant tissues by means of a biolistic particle delivery system will lead to advancements in plant genomic manipulation applications.

Delivery of biomolecules mediated by mesoporous silica nanoparticleis (MSNs) particularly interesting because without the assistance of protein transport systems, proteins are often unable to cross the barrier of cells membrane [33].

In general, the low activity of endocytosis and the barrier of cell walls limit the application of nanotechnology on plant systems. Up to now, only calcium phosphate nanoparticles and poly dendrimer (DNA vector),showed that particles could be entered into plants through simple co-culture methods and works as carriers for gene delivery without any additional assistance, and high transient delivery efficiency could be obtained [34]. However, this method was cost effective and should be developed further.

More recently, to develop plant transient gene expression system, Mou, Chang [35] used functionalized mesoporous silica nanoparticles. In this approach, MSNs worked as a cargo for deliver foreign DNA into Arabidopsis thaliana roots without any need for mechanical forces. In addition, gene expression was detected in the epidermal layer and in the more inner cortical and endodermal root tissues by both antibody labelling and fluorescence. This in particular is an advantage to the common particle bombardment technique which hardly delivers DNA into deeper plant tissues [35] (For more details, see Figure 5 and Figure 6).

3.2. Silicon Carbide-Mediated Transformation

The use of silicon-mediated transformation first reported by Kaeppler, Somers [36] which is one of the least complicated techniques. In this technique, small needle-type silicon carbide whiskers with very high tensile strength are mixed with the gene of interest and plant cells, and the mixture is vortexed and then plated on culture medium [36]. Subsequently, the cultured cells will be assessed for DNA integration and insertion into the recipient cells. In this process, the created holes permit the DNA to be delivered into plant cells [37]. Stably transformed plant cells have been produced by using this kind of DNA delivery method [36] [38]. In addition, by exposing cells to high molarity of sorbitol, the efficiency of DNA delivery by silicon carbide whiskers could be increased [39].

Table 2. Stable transformation efficiency for three coating treatments, same letters indicate lack of significant difference at $p < 0.05$, (*) For each bombardment (20 - 30) calli were placed in the centre of the petri dish to cover a circular area of 4.6 cm^2 [31].

Treatment	Experiment	Number of bombardments	Number of transgenic plants	Number of transgenic plants/bombardment*
Spermidine	I	8	13	1.63
	II	8	15	1.88
	III	14	25	1.79
Spermidine total		30	53	1.76 ± 0.07
Protamine	I	8	23	2.88
	II	8	9	1.13
	III	14	16	1.14
Protamine total		30	45	1.71 ± 0.58
Seashell DNA del TM	I	8	7	0.88
	II	8	9	1.13
	III	14	25	1.79
Seashell DNA del TM		30	41	1.26 ± 0.27

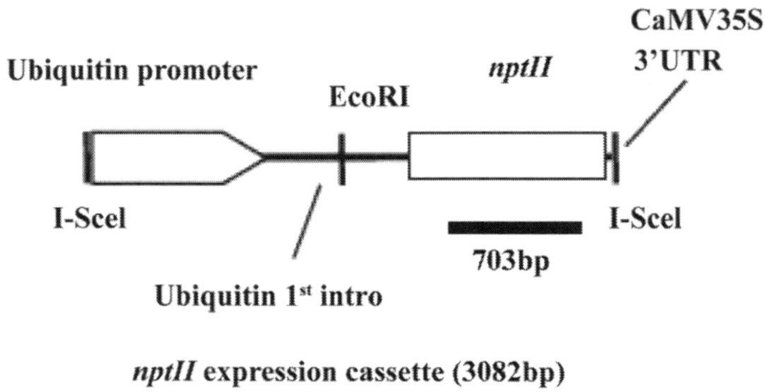

Figure 4. Schematic representation of the nptII expression cassette for stable transformation analysis. The black line underneath nptII gene represents the nptII ORF region (703 bps) that was used as probe in southern blot hybridization [31].

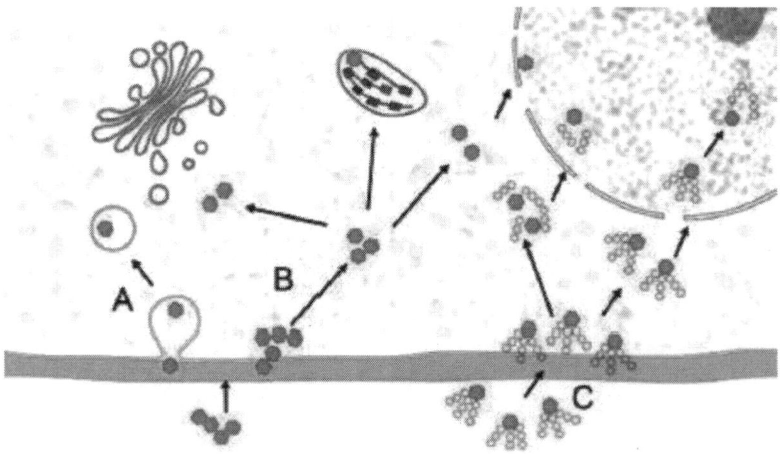

Figure 5. Possible routes of TMAPS/F-MSNs plasmid after internalization into the Arabidopsis root cell. Once passed through the cell wall, TMAPS/F-MSNs may be internalized by endocytosis (scheme A) or penetrate the plasma membrane (scheme B), which may be a primary route. Particles could enter organelles or stay in the cytoplasm after TMAPS/F-MSNs pass the plasma membrane such as plastids and the nucleus. The DNA-loaded TMAPS/F-MSN complex internalized into the plant cell (scheme C) could then reach the nucleus. MSNs and plasmid DNA may pass through the nuclear pore as a complex or DNA molecules may be released from the MSNs and enter the nucleus [3].

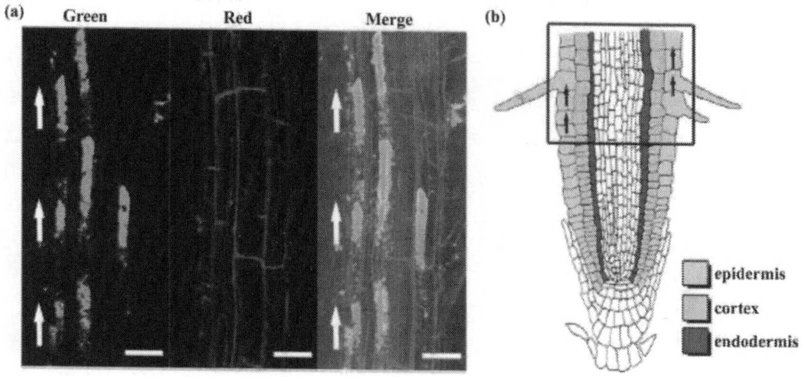

Figure 6. Polar distribution of internalized MSNs in Arabidopsis root cells. (a) Confocal images of uneven distribution of TMAPS/F-MSNs (green) in individual cells (arrows) after incubation for 24 h at 24c^0 in 1/2 MS. To label the cell walls and reveal cell viability, Arabidopsis roots were stained with PI (red). Scale bars: 50 mm. (b) a schematic illustration of the Arabidopsis root. And then in the maturation zone of the root (black box), the TMAPS/F-MSNs was detected [3].

Although the attributes of this method have been recognized for being simple, less resource requiring for DNA delivery into plants cells and cost effective [37] [38], it has also been reported that it imposes health hazards due to fiber inhalation and may be harmful to human beings, if not performed properly [40]. Therefore, there has been a search for other possible alternative, which could be used for similar DNA transfer purposes, as a consequence; the use of aluminum borate whiskers (ABW) was appeared [41]. The same group researchers improved this method by using multidirectional shaker instead of ordinary vortex. Recently, an efficient transgenic kabocha squash plants (Cucurbita moschata) have also been produced which was about (2.7%) by penetrating cotyledonary node explants with (ABW) prior to inoculation with Agrobacterium [42].

Moreover, by group of researchers in Pakistan [43], the newly inexpensive and efficient protocol was developed for salt tolerance in cotton due to transgene overexpression of Arabidopsis vacuolar pyrophosphatase (AVP1) genes by using a silicon carbide whisker-mediated gene transfer system. In addition, from just (1 g) embryogenic cotton calli treated with whiskers, they generated about 60 fertile plants of 20 independent transgenic events. Furthermore, for optimum efficiency of cotton embryogenic callus, transformations of cotton calli were evaluated giving about (94%).

3.3. Electroporation-Mediated Transformation

The application of electroporation has a major impact on genetic manipulation of organism. In this method, the foreign DNA migrates through high voltage resulting in formation of pores in the plasma membrane [44]. In general, this method was originally developed for protoplast transformation but has subsequently been revealed to work with whole plant cells [45] [46].

Practically, high electric fields are applied to protoplasts or cells suspended in a liquid culture medium enclosed in a discharge chamber [47]. Then, by a capacitor discharge, the electric field is applied which in some commercial instruments can provide high voltage, up to (3000) [48]. The voltage and time applied relies on the cell type used for electroporation. In addition, efficiency of electroporation may be affected by surface concentration of DNA and tolerance of cells to membrane permeation [49] [50].

However, there was an attention in bypassing the protoplasts for electroporation because the regeneration of fertile transgenic plants system has not been developed in most plant species [49]. In addition, the main physical barrier to electroporation maybe is the thickness of recipient cell walls [8].

The first stable transformation through this method was with rice seeds cultured (2) days prior which has been subjected by electro-injection with a plasmid having neomycin phosphotransferase II (NPTII) gene and then transgenic rice were regenerated through somatic embryogenesis [51]. Subsequently, by using the pulse-dis- charging electroporation system, stable transformants has been produced for immature corn embryos [52].

4. APPLICATION OF DISEASE RESISTANCE IN CROPS

4.1. Agrobacterium-Mediated Gene Transfer

Recently, resistance against Leaf Spot under improved form of CaMV 35S promoter by over expressing rice chitinase-3 gene in transgenic peanut through Agrobacterium mediated gene transfer was found [53]. Agrobacterium tumefaciens strain LB4404 was used holding the binary vector (pB1333-EN4-RCG3) comprising the hygromycin resistance (hpt) and chitinase gene as selectable marker (see Figure 7). In addition, through

Southern hybridization analysis of the genomic DNA, the integration of transgene in the nuclear genome of the putative transformed plants was confirmed.

More than 60% of regenerated plantlets were observed whereas transgenic healthy plants with over 42% transformation frequency were formed and all plants were normally grown. Subsequently, by infection with the microspores, (T1) plants were tested for resistance against Leaf Spot. Transgenic strains showed a higher resistance than the non-transgenic plants. Their results indicated that the lines with high enzyme activity similarly showed a high resistance against leaf spot (see Table 3).

4.2. Microprojectile Bombardment

Tissue culture and transformation were the most difficult task for developing the transgenic papaya [54]. However, this mission was taken into consideration by Fitch and Manshardt [55], three different cultivars (Sunset, Sunrise Kapoho) were targeted. To develop a papaya regeneration system through organogenesis, a great effort has been made by Fitch but unfortunately he was unsuccessful. Soon after, a quick decision has been made to use embryogenic tissue transformation through particle bombardments. Therefore, with tungsten particles coated with DNA of the PRSV HA 5-1 coat protein gene with the gene gun, embryogenic tissue was bombarded [56]. Then after 15 months in the greenhouse, transgenic plants were obtained and growing in the good conditions [57].

However, genetic engineering through particle bombardment and Agrobacterium-mediate both create relatively random integration of the transgene into the host genome [58]. In addition, the expression of the transgenes considerably affected by the random integration and similarly gene disruption could possibly occur due to this integration [58]. Southern blot hybridization, however, is useful mechanism for evaluating transgene copy number of insertion but the physical location of the transgene insertion cannot be indicated [58].

Researchers were thinking about more attractive technique to control disease resistance for papaya. Therefore, genetic resistance within the species may be more explained with the advantage of fully first sequenced genome of transgenic papaya [59]. Recently from a wild relative by back crossing, the introgression of papaya ringspot virus (PRSV) resistance was successfully accomplished [60]. This provides new hope that other wild relative traits may be introgressed into papaya in the near future.

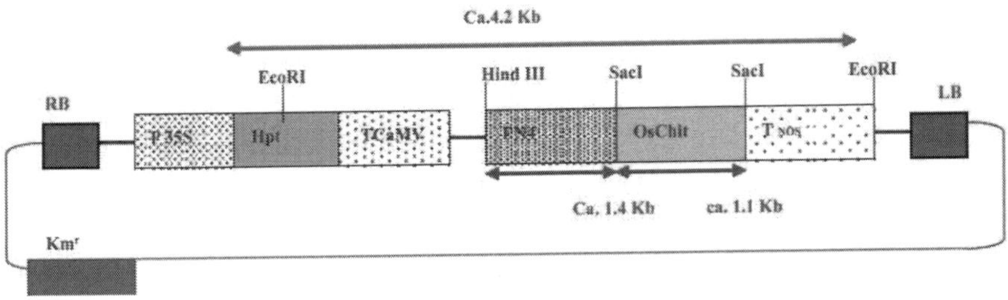

Figure 7. Vector pB1333-EN4-RCG3 [2].

5. CONCLUSIONS

Development of gene transfer systems in plants probably one of the most challenging aspects of plant research. Now day, the two methods of choice are definitely the Agrobacterium-mediated and the biolistic-mediated DNA delivery systems. From two decades ago, development of transformation technology was viewed primarily as an objective to the production of transgenic crops with improved agronomic characteristics for enhanced crop productivity [6]. However, the emergence of functional genomics and the requirement for high-throughput technology for assessment of gene function in plants have generated a whole new focus on generation of transgenic plants [61]. This circumstance has become even more evident for economically important crop plants, in which high efficiency transformation systems are generally missing, compared to model systems such as Arabidopsis thaliana [62]. However, progress made in rice and corn is encouraging with the use of Agrobacterium for transformation in a fairly high-throughput manner [63] [64]. The focus is, therefore, still focused around those other economically important plants, which can be made amenable for high-throughput gene transfer in order to make greatest use of the huge repertoire of data flowing from genome projects.

Table 3. Rice chitinase activity and pathogen resistance assay in peanut. In response to the hydrogen peroxide, Oschit transcripts accumulate. Therefore, an Oschit band with the expected size of (750-bp) has been amplified from transgenic lines [35].

Transgenic line	RT-PCR result	*Oschit* activity	Enhanced *chitinase* activity	Resistance to *C. arachidicola*	Correlation to hypothesis
Golden untransformed	–	1	–	–	–
09-G1	+	6.5	Y	Y	√
09-G2	+	3.5	Y	Y	√
09-G7	+	6.2	Y	Y	√
09-G7.3	+	3.2	Y	N	×
09-G9	+	1.9	N	N	√
09-B4	+	2.1	N	Y	×
09-B5.4	+	6	Y	Y	√
09-B13	+	2.3	Y	N	×
10-B2	+	1.8	N	N	√
10-B3.4	+	3	Y	N	×

Therefore, gene transfer methods to plants will continue to receive renewed interest in the future. The development of nanoparticles for DNA delivery into plant cells is emerging [65] and there are possibilities to combine the benefits of biolistic and Agrobacterium in the near future [66]. Meanwhile, the approach of in plant transformation using Agrobacterium is likely to continue getting attention, due to avoidance of a tissue culture step for the regeneration of transformation events. Finally, the likely achievements will come from a combination of several existing methods, with the most suitable, and most likely popular, gene transfer method having the accessibility, and simplicity for a large number of plant species.

ACKNOWLEDGEMENTS

It is a great pleasure that we can take this opportunity to send thanks and appreciation to Dr. Nawroz Abdul Razzak who were available every times and supported us by providing valuable advices. We would also like to point out the contributing role that the University of Sulaimani has played in establishing and growing the knowledge of all lecturers and supported us all of the times. Finally, we would like to express our gratitude for the sacrifices of Peshmarga and their bravery in the fight against IS.

REFERENCES

1. Kole, C. (2010) Transgenic Crop Plants: Volume 1: Principles and Development. Springer.

2. Larkin, P.J. and Scowcroft, W. (1981) Somaclonal Variation—A Novel Source of Variability from Cell Cultures for Plant Improvement. Theoretical and Applied Genetics, 60, 197-214.

3. Budar, F., et al. (1986) Agrobacterium-Mediated Gene Transfer Results Mainly in Transgenic Plants Transmitting T-DNA as a Single Mendelian Factor. Genetics, 114, 303-313.

4. Bhojwani, S.S. and Dantu, P.K. (2013) Plant Tissue Culture: An Introductory Text. Springer, London.

5. Anami, S., et al. (2013) Higher Plant Transformation: Principles and Molecular Tools. The International Journal of Developmental Biology, 57, 483-494.

6. Davey, M.R., et al. (2010) Generation and Deployment of Transgenic Crop Plants: An Overview, in Transgenic Crop Plants. Springer, 1-29.

7. Meyers, B., et al. (2010) Nuclear and Plastid Genetic Engineering of Plants: Comparison of Opportunities and Challenges. Biotechnology Advances, 28, 747-756.

8. Barampuram, S. and Zhang, Z.J. (2011) Recent Advances in Plant Transformation, in Plant Chromosome Engineering. Springer, 1-35.

9. Braun, A.C. and Laskaris, T. (1942) Tumor Formation by Attenuated Crown-Gall Bacteria in the Presence of Growth-Promoting Substances. Proceedings of the National Academy of Sciences of the United States of America, 28, 468.

10. Fraley, R.T., et al. (1985) The SEV System: A New Disarmed Ti Plasmid Vector System for Plant Transformation. Nature Biotechnology, 3, 629-635.

11. Tzfira, T., et al. (2004) Agrobacterium T-DNA Integration: Molecules and Models. Trends in Genetics, 20, 375-383.

12. Stachel, S.E. and Nester, E.W. (1986) The Genetic and Transcriptional Organization of the Vir Region of the A6 Ti Plasmid of Agrobacterium tumefaciens. The EMBO Journal, 5, 1445-1454.

13. Merlo, D.J., Nutter, R.C., Montoya, A.L., et al. (1980) The Boundaries and Copy Numbers of Ti Plasmid T-DNA Vary in Crown Gall Tumors. Molecular and General Genetics MGG, 177, 637-643.

14. Gelvin, S.B. (2000) Agrobacterium and Plant Genes Involved in T-DNA Transfer and Integration. Annual Review of Plant Biology, 51, 223-256.

15. Sone, T., Nagamori, E., Ikeuchi, T., et al. (2002) A Novel Gene Delivery System in Plants with Calcium Alginate Micro-Beads. Journal of Bioscience and Bioengineering, 94, 87-91.

16. Wada, N., et al. (2011) A Novel Gene Delivery System in Plants with Calcium Alginate Micro-Beads. In: Stewart, C.N., et al., Eds., Plant Transformation Technologies, Wiley-Blackwell, Oxford, 73-82.

17. Nakano, A., Suzuki, G., Yamamoto, M., et al. (2005) Rearrangements of Large-Insert T-DNAs in Transgenic Rice. Molecular Genetics and Genomics, 273, 123-129.

18. Chetty, V.J., Ceballos, N., Garcia, D., et al. (2013) Evaluation of Four Agrobacterium tumefaciens Strains for the Genetic Transformation of Tomato (Solanum lycopersicum L.) Cultivar Micro-Tom. Plant Cell Reports, 32, 239-247.

19. Dan, Y.H., Yan, H., Munyikwa, T., et al. (2006) MicroTom—A High-Throughput Model Transformation System for Functional Genomics. Plant Cell Reports, 25, 432-441.

20. Gardner, R.C. (1993) Gene Transfer into Tropical and Subtropical Crops. Scientia Horticulturae, 55, 65-82.

21. Sanford, J.C. (1988) The Biolistic Process. Trends in Biotechnology, 6, 299-302.

22. Kikkert, J.R. (1993) The Biolistic® PDS-1000/He Device. Plant Cell, Tissue and Organ Culture, 33, 221-226.

23. Klein, T.M., et al. (2011) Particle Bombardment: An Established Weapon in the Arsenal of Plant Biotechnologists. In: Stewart, C.N., et al., Eds., Plant Transformation Technologies, Wiley-Blackwell, Oxford.

24. Morikawa, H., Iida, A. and Yamada, Y. (1989) Transient Expression of Foreign Genes in Plant Cells and Tissues Obtained by a Simple Biolistic Device (Particle-Gun). Applied Microbiology and Biotechnology, 31, 320-322.

25. Oard, J.H., Paige, D.F., Simmonds, J.A. and Gradziel, T.M. (1990) Transient Gene Expression in Maize, Rice, and Wheat Cells Using an Airgun Apparatus. Plant Physiology, 92, 334-339.

26. Iida, A., Seki, M., Kamada, M., et al. (1990) Gene Delivery into Cultured Plant Cells by DNA-Coated Gold Particles Accelerated by a Pneumatic Particle Gun. Theoretical and Applied Genetics, 80, 813-816.

27. Taylor, N.J. and Fauquet, C.M. (2002) Microparticle Bombardment as a Tool in Plant Science and Agricultural Biotechnology. DNA and Cell Biology, 21, 963-977.

28. Bower, R. and Birch, R.G. (1992) Transgenic Sugarcane Plants via Microprojectile Bombardment. The Plant Journal, 2, 409-416.

29. Raghuwanshi, A. and Birch, R.G. (2010) Genetic Transformation of Sweet Sorghum. Plant Cell Reports, 29, 997-1005.

30. Birch, R. and Maretzki, A. (1993) Transformation of Sugarcane. In: Bajaj, Y.P.S., Ed., Plant Protoplasts and Genetic Engineering IV, Springer, Berlin, 348-360.

31. Xiong, Y., Jung, J.H., Zeng, Q.C., et al. (2013) Comparison of Procedures for DNA Coating of Micro-Carriers in the Transient and Stable Biolistic Transformation of Sugarcane. Plant Cell, Tissue and Organ Culture (PCTOC), 112, 95-99.

32. Wu, J., Du, H.W., Liao, X.W., et al. (2011) An Improved Particle Bombardment for the Generation of Transgenic Plants by Direct Immobilization of Relleasable Tn5 Transposases onto Gold Particles. Plant Molecular Biology, 77, 117-127.

33. Wu, P.P., Bai, P., Lei, Z.B., et al. (2011) Gold Nanoparticles Supported on Functionalized Mesoporous Silica for Selective Oxidation of Cyclohexane. Microporous and Mesoporous Materials, 141, 222-230.

34. Naqvi, S., Maitra, A.N., Abdin, M.Z., et al. (2012) Calcium Phosphate Nanoparticle Mediated Genetic Transformation in Plants. Journal of Materials Chemistry, 22, 3500-3507.

35. Mou, C.-Y., et al. (2013) A Simple Plant Gene Delivery System Using Mesoporous Silica Nanoparticles as a Carrier. Journal of Materials Chemistry B, 1, 5279.

36. Kaeppler, H.F., Somers, D.A., Rines, H.W. and Cockburn, A.F. (1992) Silicon Carbide Fiber-Mediated Stable Transformation of Plant Cells. Theoretical and Applied Genetics, 84, 560-566.

37. Songstad, D.D., Somers, D.A. and Griesbach, R.J. (1995) Advances in Alternative DNA Delivery Techniques. Plant Cell, Tissue and Organ Culture, 40, 1-15.

38. Wang, K., Drayton, P., Frame, B., et al. (1995) Whisker-Mediated Plant Transformation: An Alternative Technology. In Vitro Cellular & Developmental Biology-Plant, 31, 101-104.

39. Wang, J.H., Wieser, H., Pawelzik, E., et al. (2005) Impact of the Fungal Protease Produced by Fusarium culmorum on the Protein Quality and Breadmaking Properties of Winter Wheat. European Food Research and Technology, 220, 552-559.

40. Svensson, I., Artursson, E., Leanderson, P., et al. (1997) Toxicity in Vitro of Some Silicon Carbides and Silicon Nitrides: Whiskers and Powders. American Journal of Industrial Medicine, 31, 335-343.

41. Mizuno, K., Takahashi, W., Ohyama, T., et al. (2004) Improvement of the Aluminum Borate Whisker-Mediated Method of DNA Delivery into Rice Callus. Plant Production Science, 7, 45-49.

42. Nanasato, Y., Konagaya, K.-I., Okuzaki, A., et al. (2011) Agrobacterium-Mediated Transformation of Kabocha Squash (Cucurbita moschata Duch) Induced by Wounding with Aluminum Borate Whiskers. Plant Cell Reports, 30, 1455-1464.

43. Asad, S., Mukhtar, Z., Nazir, F., et al. (2008) Silicon Carbide Whisker-Mediated Embryogenic Callus Transformation of Cotton (Gossypium hirsutum L.) and Regeneration of Salt Tolerant Plants. Molecular Biotechnology, 40, 161-169.

44. Hiromichi, M., Asako, I., Chiaki, M., et al. (1986) Gene Transfer into Intact Plant Cells by Electroinjection through Cell Walls and Membranes. Gene, 41, 121-124.

45. Neumann, E. and Rosenheck, K. (1972) Permeability Changes Induced by Electric Impulses in Vesicular Membranes. The Journal of Membrane Biology, 10, 279-290.

46. Fromm, M., Taylor, L.P. and Walbot, V. (1985) Expression of Genes Transferred into Monocot and Dicot Plant Cells by Electroporation. Proceedings of the National Academy of Sciences of the United States of America, 82, 5824-5828.

47. Mischke, S., Saunders, J.A. and Owens, L. (1986) A Versatile Low-Cost Apparatus for Cell Electrofusion and Other Electro-physiological

Treatments. Journal of Biochemical and Biophysical Methods, 13, 65-75.

48. BIO-RAD (2013) Gene Pulser Xcell™ Electroporation Systems. http://www.bio-rad.com/en-us/product/gene-pulser-xcell-electropo-ration-systems?WT.srch=1&WT.mc _id=aw-gxd-eu-electroporation& WT.knsh_id=779cdfd1-7c5a-68e8-ca52-0000241ad3bd&gclid= CPbky 6CYzroCFdPltAodQHAAxg

49. Saunders, J.A., Matthews, B.F. and Miller, P.D. (1989) Plant Gene Transfer Using Electrofusion and Electroporation. In: Neumann, E., Sowers, A.E. and Jordan, C.A., Eds., Electroporation and Electrofusion in Cell Bi-ology, Plenum Press, New York, 343-354.

50. Fisk, H.J. and Dandekar, A.M. (2004) Electroporation: Introduction and Expression of Transgenes in Plant Protoplasts. In: Dunwell, J.M. and Wetten, A.C., Eds., Transgenic Plants: Methods and Protocols, Springer, New York, 79-90.

51. Xu, X., Shi, H. and Li, B. (1995) Transgenic Indica Rice Plants by Electroporation of Seed Embryo Cells. In: Potrykus, I. and Spangenberg, G., Eds., Gene Transfer to Plants, Springer, Berlin Heidelberg, 201-207.

52. Ke, X.Y., Zhang, X.W., Shi, H.P. and Li, B.J. (1996) Electroporation of Immature Maize Zygotic Embryos and Regeneration of Transgenic Plants. Transgenic Research, 5, 219-221.

53. Iqbal, M.M., Nazir, F., Ali, S., et al. (2012) Over Expression of Rice Chitinase Gene in Transgenic Peanut (Arachis hypogaea L.) Improves Resistance against Leaf Spot. Molecular Biotechnology, 50, 129-136.

54. Ming, R.M. and Paul, H. (2013) Genetics and Ge-nomics of Papaya. Springer, New York.

55. Fitch, M.M. and Manshardt, R.M. (1990) Somatic Embryogenesis and Plant Regeneration from Immature Zygotic Embryos of Papaya (Carica papaya L.). Plant Cell Reports, 9, 320-324.

56. Fitch, M.M., Manshardt, R.M., Gonsalves, D., et al. (1990) Stable Transformation of Papaya via Microprojectile Bombardment. Plant Cell Reports, 9, 189-194.

57. Fitch, M.M., Manshardt, R.M., Gonsalves, D., et al. (1992) Virus Resistant Papaya Plants Derived from Tissues Bombarded with the Coat Protein Gene of Papaya Ringspot Virus. Bio/Technology, 10, 1466-1472.

58. VanBuren, R. and Ming, R. (2014) Sequencing and Assembly of the Transgenic Papaya Genome. In: Ming, R. and Moore, P.H., Eds., Genetics and Genomics of Papaya, Springer, New York, 187-203.

59. Ming, R., Hou, S.B., Feng, Y., et al. (2008) The Draft Genome of the Transgenic Tropical Fruit Tree Papaya (Carica papaya Linnaeus). Nature, 452, 991-996.

60. Siar, S.V., Beligan, G.A., Sajise, A.J.C., et al. (2011) Papaya Ringspot Virus Resistance in Carica papaya via Introgression from Vasconcellea quercifolia. Euphytica, 181, 159-168.

61. Li, J.R., Ye, X.G., An, B.Y., et al. (2012) Genetic Transformation of Wheat: Current Status and Future Prospects. Plant Biotechnology Reports, 6, 183-193.

62. Gelvin, S.B. (2010) Plant Proteins Involved in Agrobacterium-Mediated Genetic Transformation. Annual Review of Phytopathology, 48, 45-68.

63. Ombori, O., Muoma, J.V.O. and Machuka, J. (2013) Agrobacterium-Mediated Genetic Transformation of Selected Tropical Inbred and Hybrid Maize (Zea mays L.) Lines. Plant Cell, Tissue and Organ Culture (PCTOC), 113, 11-23.

64. Saika, H. and Toki, S. (2010) Mature Seed-Derived Callus of the Model Indica Rice Variety Kasalath Is Highly Competent in Agrobacterium-Mediated Transformation. Plant Cell Reports, 29, 1351-1364.

65. Hao, Y.Z., Yang, X.Y., Shi, Y.Z., et al. (2013) Magnetic Gold Nanoparticles as a Vehicle for Fluorescein Isothiocyanate and DNA Delivery into Plant Cells. Botany, 91, 457-466.

66. Jackson, M.A., Anderson, D.J. and Birch, R.G. (2013) Comparison of Agrobacterium and Particle Bombardment Using Whole Plasmid or Minimal Cassette for Production of High-Expressing, Low-Copy Transgenic Plants. Transgenic Research, 22, 143-151.

Chapter 4

Efficacy of plant extracts in plant disease management

Malkhan Singh Gurjar[1], Shahid Ali[2*], Masood Akhtar[2], Kangabam Suraj Singh[1]

[1]Central Potato Research Station, Shillong, India

[2]SNPG College, Azamgarh (VBS Purvanchal University, Jaunpur), India

ABSTRACT

The overzealous and indiscriminate use of most of the synthetic fungicides has created different types of environmental and toxicological problems. Recently, in different parts of the world, attention has been paid towards exploitation of higher plant products as novel chemotherapeutants in plant protection. The popularity of botanical pesticides is once again increasing and some plant products are being used globally as green pesticides. Pyrethroids and neem products are well established commercially as botanical pesticides and recently some essential oils of higher plants have also been used as antimicrobials against storage pests because of their relatively safe status and wide acceptance by the consumers. Some of the volatile oils, which often contain the principal aromatic and flavouring components of herbs and spices, have been recommended as plant based antimicrobials to retard microbial contamination and reduction in spoilage of food commodities. In the context of agricultural pest management, botanical pesticides are best suited for use in organic food production in industrialized countries but can play a much greater role in the production and post harvest protection of food products in developing countries.

KEYWORDS

Botanical Pesticides; Plant Protection; Chemotherapeutants; Organoleptic

1. INTRODUCTION

The ultimate aim of recent research in this area has been the development of alternative control strategies to reduce dependency on synthetic fungicides. Plants have ability to synthesize aromatic secondary metabolites, like phenols, phenolic acids, quinones, flavones, flavonoids, flavonols, tannins and coumarins [1]. The components with phenolic structures, like carvacrol, eugenol, and thymol, were highly active against the pathogen. These groups of compounds show antimicrobial effect and serves as plant defence mechanisms against pathogenic microorganisms [2]. The volatile antimicrobial substance allicin (diallyl thio sulphinate) is synthesized in garlic when the tissues are damaged and the substrate alliin (S-allyl-L-cysteine Sulphoxide) mixes with the enzyme alliin-lyase. Allicin is readily membrane-permeable and undergoes thiol-disulphide exchange reactions with free thiol groups in proteins. Allicin effectively controlled seed-borne Alternaria spp. in carrot, Phytophthora leaf blight of tomato and tuber blight of potato as well as Magnaporthe on rice and downy mildew of Arabidopsis thaliana [3]. Application of plant products especially essential oils is a very attractive method for controlling post harvest diseases. Essential oil extracted from lemon grass (Cymbopogon spp.) post harvest anthracnose of mango fruit [4]. The anti viral protein (AVP) extracts from Bougainvillea spectabilis and Prosopis chilinesis were found to be effective in reducing the sunflower necrosis virus (SFNV) infection both in cowpea and sunflower plants [5]. At present, scientists are investigating for plant products of antimicrobial properties. It would be advantageous to standardize methods of extraction and in vitro antimicrobial efficacy testing so that the search for new biologically active plant products could be more systematic. Thousands of phyto chemicals which have inhibitory effects on all types of microorganisms in vitro should be subjected in vivo testing to evaluate the efficacy in controlling the incidence of diseases in crops, plants, and humans [2].

2. BOTANICALS

Some plant contains components that are toxic to pathogens. When extracted from the plant and applied on infested crops, these components are called botanical pesticides or botanicals.

2.1 Commonly used botanicals

Plant extracts: Neem (Azadirachta indica, A. Juss), Garlic (Allium sativum, Linn., Eucalyptus (Eucalyptus globulus, Labill., Turmeric (Curcuma Longa, Linn., Tobacco (Nicotiana tabacum, Linn., Ginger (Zingiber officinale, Rosc.

Essential oils: Nettle oil (Urtica spp.), Thyme oil (Thymus vulgaris, Linn.), Eucalyptus oil Eucalyptus globulus, Labill. Rue oil (Ruta graveolens, Linn.), Lemon grass oil (Cymbopogon flexuosus (Steud.) Wats. and Tea tree oil (Melaleuca alternifolia).

Gel and latex: Aloe vera (Tourn. Ex Linn.).

2.2 Why Consider Botanicals?

- Sustainable solutions in agriculture
- Reduce crop losses
- Eco-friendly
- Easily bio-degradable
- Organic farming
- Cheaper
- Integrated Diseases Management

3. PLANT EXTRACTS

3.1. Methods of Plant Extract Preparation

Extraction methods involve separation of medicinally active fractions of plant tissue from inactive/inert components by using selective solvents and extraction technology (**Table 1**). Solvents diffuse into the solid plant tissues and solubilize compounds of similar polarity. Quality of plant extract depends on plant material, choice of solvents and the extraction methods.

3.2. Plant Material

Plants are collected either randomly or by following leads supplied by local healers in geographical areas where the plants are found. Fresh or dried plant materials can be used as a source for the extraction of secondary plant components. Many authors had reported about plant extract preparation from the fresh plant tissues [6]. The logic behind this came from the ethno medicinal use of fresh plant materials among the traditional and tribal people. But as many plants are used in the dry form (or as an aqueous extract) by traditional healers and due to differences in water content within different plant tissues, plants are usually air dried to a constant weight before extraction. Other researchers dry the plants in the oven at about 40°C for 72 h [7]. In most of the reported works, underground parts (root, tuber, rhizome, bulb etc.) of a plant were used extensively compared with other above ground parts in search for bioactive compounds possessing antimicrobial properties.

3.3. Choice of Solvents

Successful determination of biologically active compound from plant material is largely dependent on the type of solvent used in the extraction procedure. Properties of a good solvent in plant extractions include low toxicity, ease of evaporation at low heat, promotion of rapid physiologic absorption of the extract, preservative action and inability to cause the extract to complex or dissociate. As the end product in extraction will contain traces of residual solvent, the solvent should be non-toxic and should not interfere with the bioassay [8]. The choice will also depend on the targeted compounds to be extracted. Initial screening of plants for possible antimicrobial activities typically begins by using the crude or alcohol extractions and can be followed by various organic solvent extraction methods. Water is universal solvent, used to extract plant products with antimicrobial activity. Though traditional healers use primarily water but plant extracts from organic solvents have been found to give more consistent antimicrobial activity compared to water extract [9]. Also water soluble flavonoids (mostly anthocyanins) have no antimicrobial significance and water soluble phenolics only important as antioxidant compound a study reported that extraction of tannins and other phenolics were better in aqueous acetone than in aqueous methanol. In another study, among the twenty different solvents evaluated, chloroform was found to be the best solvent for the extraction of non-polar biological active compounds [10]. Since nearly all of the identified antimicrobial compounds from plants are aromatic or saturated organic compounds, they are most often obtained through initial ethanol or methanol extraction. Thus the most commonly used solvents for preliminary investigations of antimicrobial activity in plants are methanol, ethanol and water. The other solvents used by researchers are dicholro-methane, acetone, and hexane. Some authors use a combination of these solvents to obtain the best solvent systems for extraction. [11] examined a variety of extractants for their ability to solubilize antimicrobials from plants, rate of extraction, ease of removal, toxicity in bioassay and acetone received the highest overall rating. Though there is a wide diversification in the usage of solvents, it is necessary to focus on a standardized extraction method and solvent system for a wide variety of researchers working in diverse settings to minimize the variability in the antimicrobial efficacy reports.

3.4. Extraction Methods

Variation in extraction methods are usually depend on the length of the extraction period, solvent used, pH of the solvent, temperature, particle size of the plant tissues and the solvent-to-sample ratio. The basic principle is to grind the plant material (dry or wet) finer, which increases the surface area for extraction thereby increasing the rate of extraction. In the study by [12], 5 min extractions of very fine particles of diameter 10 µm gave higher quantities than values obtained after 24 h in a shaking machine with less finely ground material. Earlier studies reported that solvent-to-sample ratio of 10:1 (v/w) solvent to dry weight ratio has been used as ideal [13]. The extraction method that has been widely used by researchers is plant tissue homogenization in solvent [9]. Dried or wet, fresh plant parts are ground in a blender to fine particles, put in a certain quantity of solvent and shaken vigorously for 5 - 10 min or left for 24 h after which the extract is filtered. The filtrate then may be dried under reduced pressure and re-dissolved in the solvent to determine the concentration. Some researchers however centrifuged (approximately 20,000 × g, for 30 min) the filtrate for clarification of the extract. Another common method is serial exhaustive extraction which involves successive extraction with solvents of increasing polarity from a non polar (hexane) to a more polar solvent (methanol) to ensure that a wide polarity range of compound could be extracted [13]. Other researchers employ soxhlet extraction of dried plant material using organic solvent. This method cannot be used for thermolabile compounds as prolonged heating may lead to degradation of compounds.

4. ANTIMICROBIAL SECONDARY METABOLITES

Plants have limitless ability to synthesize aromatic secondary metabolites, most of which are phenols or their oxygen-substituted derivatives. Important subclasses in this group of compounds include phenols, phenolic acids, quinones, flavones, flavonoids, flavonols, tannins and coumarins (**Figure 1**). These groups of compounds show antimicrobial effect and serves as plant defense mechanisms against pathogenic microorganisms (**Table 2**). Simple phenols and phenolic acid are bioactive phytochemicals consisting a single substituted phenolic ring. Phenolic toxicity to microorganisms is due to the site(s) and number of hydroxyl groups present in the phenolic compound. Quinones are characteristically highly reactive, colored compounds with two ketone substitutions in aromatic ring. Flavones, flavonoids and flavonols are phenolic structure with one carbonyl group. They are synthesized by plants in response to microbial infection and are often found effective in vitro as antimicrobial substance against a wide array of microorganisms (**Table 3**). Tannins are polymeric phenolic substances possessing the astringent property. These compounds are soluble in water, alcohol and acetone and give precipitates with proteins. Coumarins are phenolic substances made of fused benzene and α-pyrone rings. The crude sap, volatile and essential oil extracted from whole plant or specialised plant parts like roots, stem, leaves, flowers, fruits and seeds are widely used in preparing the antimicrobial compounds which are significantly used against the different plant pathogens/diseases (**Table 4**).

Table 1. Solvents used for active component extraction [1].

Water	Ethanol	Methanol	Chloroform	Dichloro-methanol	Ether	Acetone
Tannins	Alkaloids	Terpenoids	Terpenoids	Terpenoids	Alkaloids	Flavonols
Saponins	Tannins	Saponins	Flavonoids	-	Terpenoids	-
Terpinoides	Terpinoides	Tannins	-	-	Coumarins	-
-	Flavonol	Flavones	-	-	-	-

Six broad chemical groups are as follows.

- Flavonoides and isoflavonoides
- Saponins
- Steroides
- Tannins
- Phenolic and phenolic acids—Chlrogenic acid, protocatechuic acid, ferulic acid, caffeic acid
- Coumarins and Pyrones

Table 2. Mode of action of phyto chemicals [1].

Class	Sub-class	Mechanism
Phenolics	Simple phenols	Membrane disruption, substrate deprivation
Phenolic acids	Phenolic acids	Bind to adhesins, complex with cell wall, inactivate enzymes
Terpenoids, essential oils		Membrane disruption
Alkaloids		Intercalate into cell wall
Tannins		Bind to proteins, enzyme inhibition, substrate deprivation
Flavonoids		Bind to adhesins, complex with cell wall, Inactivate enzymes
Coumarins		Interaction with eucaryotic DNA
Lectins and polypeptides		Form disulfide bridges

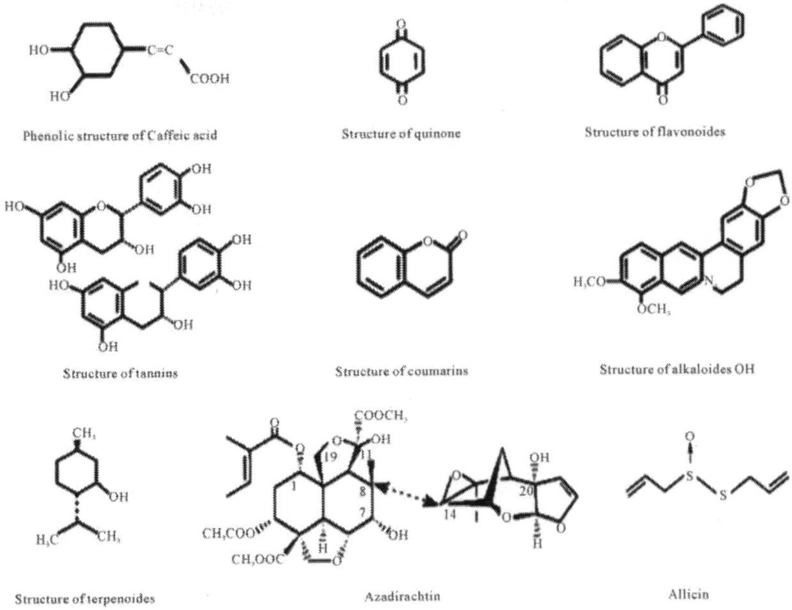

Figure 1. Structure of antimicrobial compounds [14].

Methods for Evaluation of Efficacy of Plant Extract

In vitro antimicrobial susceptibility testing (AST)

1) Diffusion test
    a) agar well diffusion
    b) agar disk diffusion
    c) poison food technique
    d) bio autography
2) Dilution methods
    a) agar dilution
    b) broth micro dilution assay
    c) broth macro dilution assay

Table 3. Botanicals produced by plants having antimicrobial activity.

Common name	Scientific name	Compound	Class	Activity
Apple	*Malus pumila* Mill.	Phloretin	Flavonoid derivative	General
Ashwagandha	*Withania somnifera* Dunal.	Withafarin A	Lactone	Bacteria, fungi
Bael tree	*Aegle marmelos* Linn.	Essential oil	Terpenoid	Fungi
Blue gum tree	*Eucalyptus globulus* Labill.	Tannin	Polyphenol	Fungi, Bacteria, Viruses
Onion	*Allium cepa* Linn.	Allicin	Sulfoxide	Fungi, Bacteria
Thyme	*Thymus vulgaris* Linn	Caffeic acid	Terpenoid	Fungi, Bacteria, viruses
Turmeric	*Curcuma longa* Linn.	Curcumin	Terpenoids	Fungi, Bacteria, protozoa
Thorn apple	*Datura stramonium* Linn.	Hyoscymine Scopolamine	Alkaloids	Fungi
Black pepper	*Piper nigrum* Linn.	Piperine	Alkaloid	Fungi
Castorbean	*Ricinus communis* Linn.	Ricinine Ricininoleic	Alkaloids	Fungi
Neem/Margosa tree	*Azadirachta indica* A.Juss.	Azadirachtin	Terpenoides	Fungi, Bacteria
Garlic	*Allium sativum* Linn.	Allicin	Solfoxide	Fungi, Bacteria

Table 4. Botanicals.

Plant	Part used	Preparations	Diseases/pathogen	References
Datura/thorn apple (*D.stamonium*) *Calotropis procera* (Ait.) R. Br. *Oscimum* spp.	Root, stem, Leaf, flowers	Crude extract	*Curvularia lunata*	[15]
Turmeric (*Curcuma longa* Linn.), Ginger (*Zingiber officinale* Rosc.)	Rhizome	Crude extract	*Phytopthora infestans, Fusarium solani, Pyricularia oryzae*	[16]
Perslane (*Portulaca olaracea* Linn.)	Leaf	Crude extract	*Helminthosporium maydis*	[17]
Hena (*Lowsonia inermis* Linn.)	Leaf	Crude extract	*Dreshslera oryzae*	[18]
Neem/Margosa (*Azadirachta indica* A.Juss.), Sugar apple (*Annona squamosa* Linn.), Holy basil (*Oscimum sanctum* Linn.)	Leaf, Stem Bark, root	Crude extract	Anthracnose of pepper	[19]
Neem/Margosa (*Azadirachta indica* A.Juss.),	Seed kernel	Oil	*A. alternata*	[20]
Ambrosioides Linn., *Oscimum* spp.	Leaf	Essential oils	*Aspergillus flavus*	[21]
Garlic (*Allium sativum* Linn.), Datura (*D. stramonium* Linn.)	Bulb, Leaf	Ethanol extracts	*Curvularia lunata*	[22]
Spearmint (*Mentha spicata* Linn.), Greek Sage (*Salvia fruticosa* Mill.), *Thymbra* spp.	Leaf	Essential oils	*Rhizoctonia solani, Sclerotium sclerotiorum*	[23]
Spanish flag (*Lantana camara* Linn.)	Leaf	Crude Extracts	Castor grey rot (*Botrytis ricini*)	[24]
Neem/Margosa (*Azadirachta indica* A.Juss.),	Seed, Leaf	Crude Extracts	Early blight of tomato	[25]
Madar (*Calotropis procera* (Ait.) R.Br.	Leaf	Crude Extracts	Tikka leaf spot disease of groundnut	[26]
Neem/Margosa (*Azadirachta indica* A.Juss.),	Seed	NSKE	Powdery mildew of pea	[27]
Spanish flag (*L.camara* Linn.), Pongam (*Pongamia pinnata* L.Pierre.)	Leaf	Crude extracts	Leaf blight of onion	[28]
Holy basil (*Oscimum sanctum*Linn.), peach (*Prunus persica* Linn.) Stokes.	Leaf	Essential oil	Grey mould (*Botrytis cinerea*) of grapes	[29]

Neem/Margosa (*Azadirachta indica* A.Juss.),	Leaf	Achook formulations (azadirachtina)	Sheath blight of rice	[30]
Neem/Margosa (*Azadirachta indica* A.Juss.),	Seed kernel	Neem oil	Rice tungro virus	[31]
Neem/Margosa (*Azadirachta indica* A.Juss.),	Leaf, Seed	Achook, Neemazal,	Bacterial blight of rice	[32]
Oregano (*Origanum hercleoticum* (weed species)	Leaf	Essential oils	*Fusarium oxysporum, Phoma tracheiphila*	[4]
Neem/Margosa (*Azadirachta indica* A.Juss.), Black cumin (*Nigelia sativa* Linn. Asfetida (*Ferula asafoetida* Linn.)	Seeds	Essential oils	*Fusarium oxysporum, A.niger, A.flavus*	[33]
Strawberry (*Fragaria* spp.)	Fruit	Volatile compounds	Anthracnose of strawberry	[34]
Raspberry (*Rubus* spp.) and Strawberry (*Fragaria* spp.)	Fruit	Volatile compounds	Post harvest decay fungi	[35]
Garden croton (*Codiaeum variegatum* Linn.)	Leaf	Phenolic compounds	*Alternaria alternate, Fusarium oxysporum*	[36]
Oleander (Nerium *oleander* Linn.)	Leaf	Crude extracts	Brown spot of rice (*Bipolaris oryzae*)	[37]
Indian aloe (*Aloe barbadensis* Mill.) Neem/Margosa (*Azadirachta indica* A.Juss.), Tobacco (*Nicotiana tabacum* Linn.)	Leaf	Crude extracts	Dry rot of yam *F. oxysporum, A.nizer*	[38]
Black pepper (*Piper nigrum* Linn., Clove (*Syzygium aromaticum* (Linn.) Merr. & Perry, Geranium (*Pelargonium graveolens* L'Herit), Nutmeg (*Myristica fragrans* Houtt.), (*Origanum vulgare* spp. *hirtum* (Link) Letsw. and thyme [*Thymus vulgaris* Linn.	Leaf	Volatile oil	Anti bacterial (gram positive and gram negative)	[39]

Metasequoia glyptostroboides	Leaf	Essential oil	*Fusarium oxysporum*, *Fusarium solani*, *Phytophthora capsici*, *Colletotrichum capsici*, *Sclerotinia sclerotiorum*, *Botrytis cinerea* and *Rhizoctonia solani*	[40]
Clove (*Syzigium aromaticum* Linn.), Turmeric (*Curcuma longa* Linn.), garlic (*Allium sativum* Linn.) and Holy Basil (*Ocimum Sanctum* Linn.)	Leaf, seed, fruit	Crude extract	*Aspergillus flavus*	[41]
Aspilia africana, *Chromolaena odorata*, *Musa paradisiaca* and *Tithonia diversifolia*	Leaf	Aqueous extract	*Cercospora* leaf spot of Seasame (*Sesamum indicum*) L.)	[42]
Lemon grass (*Cymbopogon* spp.), Thyme (*Thymus vulgaris* Linn.)	Leaf, root	Volatile compound	Black Mould Disease on Onion Bulbs (*Aspergillus niger*)	[43]
Brassica napus and Tomato (*Lycopersicon esculentum* Mill.)	Leaf, stem	Water extract	Bacterial disease on Onions	[44]
Ginger (*Zingiber officinale* Rosc.), aloe (*Aloe* vera), bitter kola (*Garcinia cola*) and neem (*Azadirachta indica* A.Juss.)	Leaf, Fruit and Seed	Crude	Root rot disease of cow pea (*Vigna unguiculata* L.)	[45]
Chloranthus japonicas, *Paulownia coreana*	Roots, stem	Crude extract	rice blast, rice sheath blight, and wheat leaf rust	[46]
Urtica dioica L., thyme *Thymus vulgaris* L., *Eucalyptus* spp., *Ruta graveolens* L. and *Achillea millefolium* L.	Leaf	Volatile oil	*Aletrnaria alternate*	[47]
Acacia nilotica, *Achras zapota*, *Datura stramonium*, *Emblica officinalis*, *Eucalyptus globules*, *Lawsonia inermis*, *Mimusops elengi*, *Peltophorum pterocarpum*, *Polyalthia longifolia*, *Prosopis juliflora*, *Punica granatum* and *Sygigium cumini*	Leaf, Seed, fruits	Aqueous extract	*Aspergillus* spp.	[48]
Eugenia aromatica , *Piper betle*), *Alpinia galanga* and *Sphaeranthus indicus*	Leaf	Crude	Stem rot disease of Vanilla	[49]
Ocimum gratissimum, *Aframomum melegueta*	Leaf	Crude	Post harvest yam (*Diascorea* spp.) rot	[50]
Neem (*Azadirachta indica* A.Juss.)	Leaf	Crude	Protozoan, Bacteria, Antifertility	[51]

5. VOLATILE OILS

- Volatiles are small molecular weight organic compounds having appreciable vapour pressure at ambient temperature
- Play role in defense system
- Inhibit the growth of pathogen
- Generally Recognized As Safe (GRAS)
- Isolated by hydro distillation or steam
- Terpinoides—monoterpenes (C10), ses quiterpenes (C15), diterpines (C20) Examples: Black pepper (Piper nigrum Linn.), Clove (Syzygium aromaticum Linn.) Merr. & Perry, Nutmeg (Myristica fragrans Houtt.), Oregano (Origanum vulgare Linn.) and Thyme (Thymus vulgaris Linn.)

6. ESSENTIAL OILS

- The potential essential oils used against post harvest pathogens.
- Fungistatic
- Terpenoides and aromatic compounds
- Component—carvacrol, thymol, cymene, terpine, phenylpropene derivatives, eucalyptol and anisol Examples: Nettle oil (Urtica spp.), Thyme oil (Thymus vulgais Linn.), Eucalyptus oil (Eucalyptus globules Labill., Rue oil (Ruta graveolens Linn.), Lemon grass oil (Cymbopogon flexuosus (Steud.) Wats.) and Tea tree oil (Melaleuca alternifolia Maiden & Betche).

7. LIMITATIONS OF BOTANICALS FOR PLANT DISEASE MANAGEMENT

- Extraction methods are not standardized

- Rapid degradation
- Most studies are in vitro efficacy
- Need the development of formulations
- Some chemical compounds are harmful to human and plants
- Less effective
- Less availability formulations

8. CONCLUSION

Plants contain thousands of constituents and are valuable sources of new and biologically active molecules possessing antimicrobial property (**Table 4**). The ethnobotanical study of plant is important for modern day medicine but its usefulness cannot be overemphasized if methods are not standardized to obtain comparable and reproducible results. At present, scientists are investigating for plant products of antimicrobial properties. It would be advantageous to standardize methods of extraction and in vitro antimicrobial efficacy testing so that the search for new biologically active plant products could be more systematic and interpretation of results would be facilitated. Thousands of phytochemicals which have inhibitory effects on all types of microorganisms in vitro should be subjected in vivo testing to evaluate the efficacy in controlling the incidence of disease in crops, plants, and humans. Efficient collaborations with pharmacologists and medical doctors, plant pathologists and microbiologists are crucial to see the complete development of an interesting lead compound into an exploitable product.

REFERENCES

1. Cowan, M.M. (1999) Plant products as antimicrobial agents. Clinical Microbiology Reviews, 12, 564-582.

2. Das, K., Tiwari, R.K.S. and. Shrivastava, D.K. (2010) Techniques for evaluation of medicinal plant products as antimicrobial agent: Current

methods and future trends. Journal of Medicinal Plants Research, 4, 104-111.

3. Slusarenko, A.J., Patel, A. and Portz, D. (2008) Control of plant diseases by natural products: Allicin from garlic as a case study. European Journal of Plant Pathology, 121, 313-322.

4. Salomone, A., Scaritto, G., Sacco, A., Cabras, G. and Angioni, A. (2008) Inhibitory effects of the main compounds of oregano essential oil against some pathogenic fungi. Modern fungicide and Antifungal compounds V. Deutsche Phytomedizinische Gesellschaft, 44, 345-360.

5. Lavanya, N., Saravanakumar, D., Rajendran, L., Ramiah, M., Raguchander, T. and Samiyappan, R. (2009) Management of sunflower necrosis virus through anti-viral substances. Archives of Phytopathology and Plant Protection, 42, 265-276.

6. Tiwari, R.K.S., Chandravanshi, S.S. and Ojha, B.M. (2005) Efficacy of extracts of medicinal plant species on growth of Sclerotium rolfsii root rot in tomato. Journal of Mycology and Plant Pathology, 34, 461-464.

7. Salie, F., Eagles, P.F.K. and Lens, H.M.J. (1996) Preliminary antimicrobial screening of four South African Asteraceae species. Journal of Ethnopharmacology, 52, 27- 33.

8. Ncube, N., Afolayan, S.A.J. and Okoh, A.I. (2008) Assessment techniques of antimicrobial properties of natural compounds of plant origin: Current methods and future trends. African Journal of Biotechnology, 7, 1797-1806.

9. Parekh, J., Jadeja, D. and Chanda, S. (2005) Efficacy of aqueous and methanol extracts of some medicinal plants for potential antibacterial activity. Turkish Journal of Biology, 29, 203-210.

10. Harmala, P., Vuorela, H., Tornquist, K. and Hiltunen, R. (1992) Choice of solvent in the extraction of Angelica archangelica roots with reference to calcium blocking activity. Planta Medica, 58, 176-183.

11. Eloff, J.N. (1998) A sensitive and quick microplate method to determine the minimum inhibitory concentration of plant extracts for bacteria. Planta Medica, 64, 711-713.

12. Eloff, J.N. (1998) Which extractant should be used for the screening and isolation of antimicrobial components from plants? Journal of Ethnopharmacology, 60, 1-8.

13. Green, R.J. (2004) Antioxidant activity of peanut plant tissues. Master Thesis, North Carolina State University.

14. Ciocan, I.D. and Băra, I. (2007) Plant products and antimicrobial agents. Universitatii ale Ştiinţifice Analele Alexandru Ioan Cuza, Tom VIII.

15. Manoharachary, C. and Gourinath, A. (1988) Effects of plant extracts on four pathogenic fungi. 5th International Congress of Plant Pathology, Kyoto.

16. Bandara, B.M.R., Kumar, N.S. and Samaranayake, K.M.S. (1989) An antifungal constituent from the stembark of Butea monosperma. Journal of Ethnopharmacology, 25, 73-75.

17. Noriel, L.M. and Robles, R.P. (1990) Fungicidal activity of Portulaca oleracea extract against Helminthosporium maydis Wisik and Miyake in corn (Zea mays L.). Philippine Journal of Weed Science, 17, 26-32.

18. Natarajan, M.R. and Lalithakumari, D. (1987) Antifungal activity of the leaf extract of Lawsonia inermis on Dreschslera oryzae. Indian Phytopathology, 40, 390-395.

19. Nduagu, C., Ekefan, E.J. and Nwankiti, A.O. (2008) Effect of some crude plant extracts on growth of Colletotrichum capsici (Synd) Butler & Bisby, causal agent of pepper anthracnose. Journal of Applied Biosciences, 6, 184-190.

20. Dharam, V. and Sharma R.K. (1985) Efficacy of fungicides XXIX studies on the fungicidal properties of neem oil. Indian Journal of Plant Pathology, 3, 241-242.

21. Mishra, A.K., Dwivedi, S.K. and Kishore, N. (1989) Antifungal activity of some essential oils. National Academy Science Letters, 12, 335-336.

22. Upadhyaya, M.R. and Gupta, R.C. (1990) Effect of extracts of some medicinal plants on the growth of Curvularia lunata. Indian Journal of Mycology and Plant Pathology, 20, 144-145.

23. Yegen, O., Begger, B. and Heitefuss, R. (1992) Studies on the fungitoxic effect for extracts of six selected plants from Turkey on phytopathogenic fungi. Zeitschrift fur Pflanzenkrankheiten und Pflanzenschütz, 99, 349-359.

24. Bhattiprolu, S.L. and Bhattiprolu, G.R. (2006) Management of castor grey rot disease using botanical and biological agents. Indian Journal of Plant Protection, 34, 101-104.

25. Patil, M.J., Ukey, S.P. and Raut, B.T. (2001) Evaluation of fungicides and botanicals for the management of early blight (Alternaria solani) of tomato. PKV Research Journal, 25, 49-51.

26. Srinivas, T., Rao, M.S., Reddy, P.S. and Reddy, P.N. (1997) Integrated management of leaf spot of groundnut (Arachis hypogaea L.) with botanicals and chemicals. Zeitschrift fur Pflanzenkrankheiten und Pflanzenschutz, 104, 528-530.

27. Surwase, A.G., Badgire, D.R. and Suryawanshi, A.P. (2009) Management of pea powdery mildew by fungicides, botanicals and bio-agents. Annals of Plant Protection Sciences, 17, 384-388.

28. Bhosale, M.V., Perne, R.R., Shete, M.H. and Pawar, N.B. (2008) Efficacy of botanicals, bioagents and fungicides on management of leaf blight (Alternaria porri) of onion. Journal of Plant Disease Sciences, 3, 88-89.

29. Tripathi, P., Dubey, N.K. and Shukla, A.K. (2008) Use of some essential oils as post harvest botanical fungicides in the management of grey mould of grapes caused by Botrytis cinerea. World Journal of Microbiology and Biotechnology, 24, 39-46.

30. Kandhari, J. (2007) Management of sheath blight of rice through fungicides and botanicals. Indian Phytopathology, 60, 214-217.

31. Muthamilan, M. and Revathy, N. (2007) Management of tungro virus disease of rice with antagonists and botanicals. Journal of Ecobiology, 19, 81-88.

32. Sunder, S., Singh, R. and Dodan, D.S. (2005) Management of bacterial blight of rice with botanical extracts and non-conventional chemicals. Plant Disease Research, 20, 12-17.

33. Sitara, U., Naseem, J. and Sultana, N. (2008) Antifungal effect of essential oils on in vitro growth of pathogenic fungi. Pakistan Journal of Botany, 40, 409-414.

34. Arroyo, F.T., Moreno, J., Daza P., Boianova, L. and Romero, F. (2007) Antifungal activity of strawberry fruit volatile compounds against Colletotrichum acutatum. Journal of Agricultural and Food Chemistry, 14, 5701- 5707.

35. Vaughan, S.F., Spenser, G.F. and Shasha, B.F. (1993) Volatile compounds from raspberry and strawberry fruit inhibit postharvest decay fungi. Journal of Food Science, 58, 793-796.

36. Naidu, G.P. (1988) Antifungal activity in Codieum variegatum leaf extract. Current Science, 57, 502-503.

37. Harish, S., Saravankumar, D. Ebenezar, E.G. and Seetharaman, K. (2008) Use plant extracts and biocontrol agents for the management of brown spot disease in rice. BioControl, 53, 555-567.

38. Taiga, A. (2009) Efficacy of selected plant extracts in the control of fungal dry rot of white yam (Dioscorea Rotundata) Tubers in Kogi State. American-Eurasian Journal of Sustainable Agriculture, 3, 310-313.

39. Dorman, H.J.D. and Deans, S.G. (2000) Antimicrobial agents from plants: Antibacterial activity of plant volatile oils. Journal of Applied Microbiology, 88, 308-316.

40. Bajpai, V.K. and Kang, S.C. (2010) Antifungal activity of leaf essential oil and extract of Metaseqoia glystostroboides Miki ex Hu. Journal of the American Oil Chemists' Society, 87, 327-336.

41. Reddy, K.R.N., Reddy, C.S. and Muralidharan, K. (2009) Potential of botanicals and biocontrol agents on growth and aflotoxin production by Aspergillus flavus infecting rice grains. Food Control, 20, 173-178.

42. Enikuomehin, O.A. (2005) Cercospora leaf spot disease management in Seasame (Seasamum indicum L.) with plant extracts. Journal of Tropical Agriculture, 43, 19-23.

43. Abd-Alla, M.A., El Mohamedi, R.S.R. and Badeaa, R.I. (2006) Effect on some volatile compounds on black mould disease on onions bulbs

during storage. Research Journal of Agriculture and Biological Sciences, 2, 384- 390.

44. Kowalska, B. and Smolinska, U. (2008) The effect of selected plant materials and extract on the development of bacterial disease on onion. Vegetable Crops Research Bulletin, 68, 33-45.

45. Suleiman, M.N. and Emua, S.A. (2009) Efficacy of four plant extract in the control of root rot disease of cow pea (Vigna unguiculata Linn. Walp). African Journal of Biotechnology, 8, 3806-3808.

46. Choi, G.J., Jang, K.S., Kim, J.S., Lee, S.W., Cho, J.Y., Cho, K.Y. and Kim, J.C. (2004) In vivo fungal activity of 57 plant extracts against six plant pathogenic fungi. Plant Pathology Journal, 20, 184-191.

47. Hadizadeh, I., Peivastegan, B. and Hamzehzarghani, H. (2009) Antifungal activity of essential oil from some medicinal plants of Iran against Aletrnarea alternata. American Journal of Applied Sciences, 6, 857-861.

48. Satish, S., Raveesha, K.A. and Janardhana, G.R. (1999) Antibacterial activities of plant extract on phytopathogenic Xanthomonos compestris pathovars. Letters in Applied Microbiology, 28, 145-147.

49. Suprapta, D.N. and Khalimi, K. (2009) Efficacy of plant extract formulation to suppress stem rot disease on vanilla seedlings. ISSA Journal, 15, 34-41.

50. Okigbo, R.N. and Ogbonnaya, U.O. (2006) Antifungal effect of two tropical plants leaf extracts (Ossimum gratissimum and Aframomum meleguata) on post harvest yam (Diascorea spp.) rot. African Journal of Biotechnology, 5, 727-731.

51. Atawodi, S.E. and Atawodi, J.C. (2009) Azadirachta indica (neem) a plant of microbial and pharmacological activity. Phytochemistry Reviews, 8, 601-620.

Chapter 5

Situation of Biofungicides Reconnaissance, a Case of Anthracnose Disease of Cowpea

Vitus Ikechukwu Obi*, Juan Jose Barriuso-Vargas

Department of Agricultural Science and Natural environment, University of Zaragoza, Zaragoza, Spain

ABSTRACT

Plant extracts have long been used in commercial agriculture as anti-microbial tools in food safety applications. These offer growers and agrobiologists many unique benefits which include their eco-friendliness. This work reviews the situation of Biofungicides reconnaissance in reference to fungal disease of cowpea. Twenty different pathogens were associated with various fungal diseases of cowpea and, only the species of Colletotrichum was found to have the virulence and propensity of afflicting a 100% infection on a single susceptible cowpea crop. Plant families under the affliction of Colletotrichum were analyzed. The different forms of botanicals so far availed for use as potential biofungicidal were identified. Eighteen plant families were found to represent the entire plants and plant materials agrobiologically screened within a range of thirteen years and found to habour large spectra of species containing substances of biofungicidal potentials. Current position in the use of Botanicals to combat agricultural pests and disease is 7% of the total cowpea disease management options.

KEYWORDS

Anthracnose; Biofungicides; Biopesticides; Colletotrichum destructivum; Cowpea.

1. INTRODUCTION

Cowpea (Vigna unguiculata (L.) Walp) is a leguminous, annual grain crop in the bean family (Fabaceae/Leguminosae), with high degree of variation in growth habit, leaf shape, flower colour and seed size and colour, cultivated mostly in the humid tropics of the globe for its seeds, as a vegetable crop, green manure, fodder, as a cash crop and or cover crop [1] -[4]. The crop's haulms are also valuable source of livestock protein [5].

Cowpea is a native to Africa where they are often intercropped with maize (Zea mays L.), Sorghum (Sorghum bicolor (L.) Moench), Pearl millet (Pennisetum americanum (L.) Leeke) and Cassava (Manihot esculenta Crantz) [6] [7]. This all important global crop, encounters a number of operational constraints, including pests and diseases that limit its production and yield potentials from seedling to harvest [3] [4] [7] and often provoking grain yield loss of over 35% [8] [9].

Major pathogenic groups associated with cowpea diseases, include: fungi, bacteria, viruses, nematodes and the parasitic flowering plants [3]. And Anthracnose is its major disease, causing severe damage and loses under low temperature, high humidity and free moisture [3] [6]. Though the major fungal pathogen of Cowpea crop is the Colletotrichum destructivum O'Gara, there are some other old and new fungal pathogens that influence its existence, reproduction and survival. Some of these pathogens attack and infect the roots of the crop [10] ; stems [11] [12] ; leaves [3] [13] [14] ; pods and fruits [15] [16] ; seeds/seedlings [17] [18] ; whole plant parts [4].

C. destructivum is polycyclic, having multiple life cycles in a growing season [19], and hemibiotrophic, thriving in both living and dead tissues [4]. The pathogen is seed borne fungus [20] [21], and can survive for at least two years on diseased stem tissues, plant debris, either on the soil surface or beneath [4].

The C. destructivum sporulates readily on infected cowpea at localized infection foci to produce anthracnose symptom within 96hrs of inoculation

[22], in the form of lesions as small angular brown spots on the leaf petiole, the lower surface of leaves and leaf veins of cowpea grown under different cropping patterns [6]. These various spots created later coalesced to produce a brick red to brown discoloration of the entire leaf. Symptoms are usually delayed (delitescent infection) until production of flowering buds. This degree of virulence on cowpea often leads to product yield loss between 35% and 50% [8] [9].

Available management techniques for Anthracnose disease of cowpea include the use of Biocontrol systems (bioagents), pesticides (conventional/ synthetic chemicals), cultural observations (clean seeds/hygienic fields and practices), HPR (host plant resistance) and botanicals (Biopesticides/no synthetic chemicals) [23]. Some, though seem effective, are enlaced with residual and often negative and indelible impressions [24].

The global quest for "back to nature" continues to augment the need and desirability to search for the alternative which employs natural agro-biological balance (Biopesticides) to address plant disease issues. This is remotely aimed at protecting the soil that supports the life of these crops among other horticultural plants [25], and in extension safeguards the environment for every other vivo organism.

The Environmental Protection Agency (EPA-USA) defines a Biopesticides as a pesticide derived from natural materials such as animals, plants, bacteria and certain minerals [26]. Biopesticides of plant origin are the botanicals (Figure 1). Biological control of plant disease through the use of antagonistic micro organism [4] [27] -[29] and botanical control of plant disease through the use of plant extracts [7] -[9] are two major ways in the control of plant disease in respect to natural agro biological balance.

In the evaluation of some botanicals against C. destructivum, Akinbode and Ikotun [4] inhibited the growth of the pathogen in vitro using Nicotiana tabacum plant extract. Crude botanical extracts from stem back and root bark of Azadiractha indica, Vernonia amygdalina and Cochlospermum planchonii exhibited strong fungi toxicity against Colletotrichum capsici as reported by Nduagu et al. [25].

Palhano et al. [30] inactivated spores of Colletotrichum gloeosporiodes using high hydrostatic pressure separate and combined with Citral or lemongrass (Cymbopogon citratus) essential oil. Their work reiterated the need for the use of plant essential oils as an alternative for crop health problems considering the safety and stability of the soil and its environment.

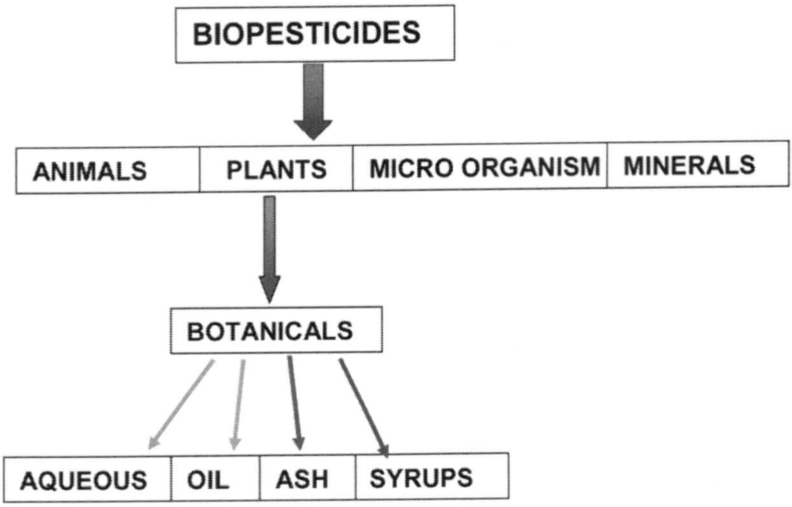

Figure 1.Botanicals in Biopesticides.

The search for bioactive substances from the plant world has led researches to distinctive regions of plants. Flowers, leaves, barks, seeds, fruits, roots and at times whole plant could be employed in the search for botanicals [31]. The seeds of neem, Azadiractha indica A. Juss and fruits of bush pepper, Xylopia aethiopica (Dunal) A. Rich were used in the work of Amadioha and Obi [8], while in another similar study the same authors employed the leaves of scent plant, Ocimum gratissimum (L.) and lemon grass, Cymbopogon citratus D. C. Stapf., for anthracnose disease of cowpea [32]. Amadioha [9] used the leaves of Piper nigrum, Ocimum sanctum and Citrus limon in his study against C. lindemuthianum.

Akinbode & Ikotun [4] and Colpas et al. [23] utilized the leaves of Nicotiana tabacum and Ricinus communis, and Ocimum gratissimum respectively in the scientific investigation for bioagents/botanicals in control of C. destructivum. Nduagu et al. [25] screened eleven plants on growth of Colletotrichum capsici (Synd) Butler & Bisby concentrating on leaves, stem bark and root barks of the concerned plants.

Different methods and techniques have been employed by scientists in the extraction and characterization of products from plants. A plant material could thus be harnessed in its fresh, or air/sun/oven-dried form, and with an adoption of extraction methods such as the use of hot or cold water for aqueous botanicals [4] [23] [25] [32] [33] ; organic solvents for oil botanicals [8] ; crude ashes for crude powder botanicals [34] [35]

There are weighty merits for the quest for wider exploration of Biopesticides in the field of agriculture. Biopesticides tend to pose fewer risks than conventional pesticides; Biopesticides are usually inherently less toxic than conventional pesticides; when used as a component of integrated pest management (IPM) programmes [25]. Biopesticides can greatly decrease the use of conventional pesticides, while crop yields remain high; Biopesticides require much less data and time frame to register than a conventional pesticide [36]. They are non residue producing control agent, making them eco-friendly and easy to use class of Reduced-risk Fungicides. It is in the support of all these that the European Community, like in other developed parts of the world, established a European Commission Working Document (SANCO/10472rev.5). This specifies data requirements for active substances of plants protection products made from plants or plant extracts.

2. DISCUSSION

The global comparison of scientific research and publications, on protection of Vigna unguiculata (L.) (Table 1), projected the Asian region with the highest of 35.9%. This was followed by Africa with 24.80%. The least scientific research and publications on protection of cowpea was from Australia with 1.40%.Scientific documents on cowpea diseases according to pathogen groups, for five years range indicated the fungi to be on top with 35.6% (table 2).This relatively high level percentage of scientific papers points at the major pathogenic constrains and its economic importance in the cultivation of cowpea crops. Regrettably the Biopesticides (Botanicals) which is a protective drive for a natural agro-biological balance in the fight against agricultural pests and disease was associated with about 7% of the total cowpea disease management options (Figure 2), a clear indication of under exploration of this area. This corroborates with the observation in the work of Emechebe & Lagoke [3].

It was observed that each cowpea pathogen has different regions of interest on a "whole cowpea plant". This work, therefore, considered a cowpea crop from different botanical dimensions of six parts, and it was discovered that 30% of the fungal infections occur on the foliar part of the crop, 25% on the stems, 15% on the roots, 10% on the pods/fruits, 25% on the seeds/seedlings and 10% on the whole parts of the plant (figure 3).

There was an indication that while the other nineteen pathogenic fungal species (table 3) poses the ability each of attacking only about a meager 20% of a cowpea crop, Colletotrichum species especially the C. destructivum & C. truncatum, have in stock 100% virulence on a single crop each at a given pathogenic situation (table 3). This corroborates with the findings of Latunde-Dada et al. [22], Latunde-Dada & Lucas [37], and Akinbode & Ikotun [4].

There are eighteen plant families presently under the anguish of the Colletotrichum Corda (table 4). About 28% family interaction existed between plant families under the affliction of Colletotrichum and the plant families screened for antifungal properties (table 6) as derived from table 4 & table 5. These apparitions were observed within the five plant families of Asteraceae, Caricaceae, Fabaceae, Lauraceae and Poaceae. The rest of about 72% were unique in occurrence, hence no family interactions among them (table 6.) The extrapolated values appear to affirm the indication on table 5 of this work that of all the entire plant families in existence only about eighteen (18) has been screened for their biofungicides characteristics between 1998 and 2011.

The products screened during this thirteen years span were of various forms or state, such as (1) Aqueous: botanicals extracted using water as the solvent. The water also forms the extract solution [4] [25] ; 2) Syrup: botanicals of a higher measure of viscosity having been extracted with a solvent other than water, and also containing some of the extracting liquid in its solution [23] [38] ; 3) Oil: botanicals in the form of essential oil of the test plants, usually extracted through a condensation system [8] ; and 4) Ash: botanicals produced in the form of residues powder left after the combustion of a test plant material [34] [35].

Table 1. Relative contributions, geographically, of scientific publication on cowpea diseases. Source: [3].

Geographical area.	Number of papers.	Percentage of papers contributed (%).
Africa	85	24.80
Asia	123	35.90
Australia	5	1.40
Europe	28	8.20
North America	75	21.80
S. America & the Caribbean	27	7.90

Table 2. Publications, for five years, on cowpea diseases according to pathogen groups. Source: [3].

Group of pathogen	Number of papers	Percentage of papers (%)
Bacteria	21	6.10
Fungi	122	35.6
Nematodes	69	20.10
Parasitic plants	19	5.50
Viruses	112	32.7

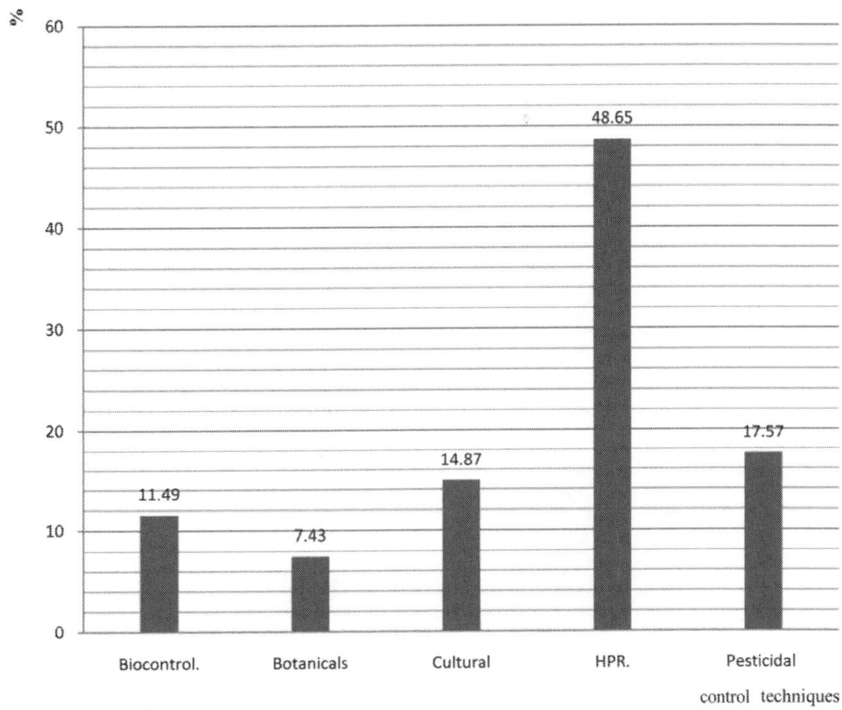

Figure 2. Percentage disease control techniques on Vigna unguiculata for a media decade.

Nevertheless, the botanical forms enumerated in this study could be extended to produce additional form by the application of further processing treatment on the original form. For example the Syrup produced in the work of Win et al. [38] was utilized in its dry crude botanical extract state after subjecting the initial extract syrup to an evapoconcentration system. This study, however, observed that most of the botanicals screened for this span of thirteen years (1998 to 2011) were produced and also utilized in their aqueous form. And considering the total 33 frequency occurrence of botanical forms (table 5), the aqueous botanicals was 51.52%, followed by the syrup (15.15%), ash (18.18%) and the oil form of 15.15%. The relative easy and economy of production could be responsible for the high percentage value obtained with aqueous botanical evaluation.

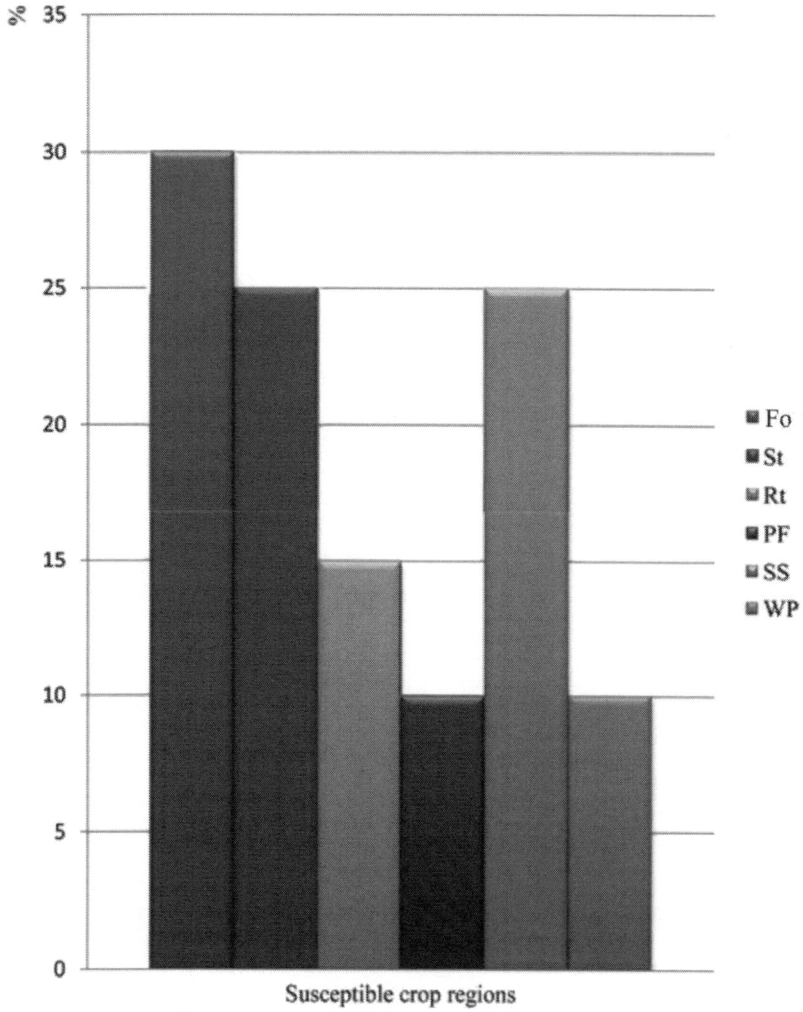

Figure 3. Percentage regional fungal attack on Vigna unguiculata crop. (Fo = foliar; St = stem; Rt = root; PF = pod/fruit; SS = seed/seedlings; WP = whole plant).

3. CONCLUSIONS

Anthracnose disease remains a devastating health problem to cowpea crop and an equated hindrance to its economic cultivation. The major afflicting pathogen Colletotrichum destructivum O'Gara has the virulence of one hundred percent (100%) infection on a single crop stand (that is every part of the crop is subject to attack and infection by C. destructivum at a given pathogenic situation).

The use of botanicals remains suitable contest to adequate disease management options, at least for its characteristics ease of production, economy and ecological amiability.

This study has availed the fact that with the high global yearning for the urgent replacement of conventional (chemical) fungicides in disease management with the ecologically compatible bio-fungicides, the seemingly several works in this direction is merely about 7% of the different management systems as indicated on cowpea disease control options and therefore, advocates for more reconnaissance in this essential area of agro-biological system whose endpoint includes the provision of an eco-friendly global environment.

Table 3. Colletotrichum Corda and other fungal pathogens of Vigna unguiculata.

Pathogen	Disease afflicted	Region affected	Reference
Alteneria cassiae Juria & Khan	*Alternaria* leaf spot	Foliar	[39]
Cercospora canescens Ellis & Martin	*Cercospora* leaf spots	Foliar	[13]
Choanephora cucurbitarum (Berk & Rav.)	*Choanephora* pod rot	Pods/fruits	[15]
Colletotrichum dematium (Pers. ex Fr.)	*Colletotrichum* stem disease	Stem	[12]
Colletotrichum destructivum O'Gara	Anthracnose	Every part	[4]
Colletotrichum capsici (Syd.) Butl. & Bisb. (=*Colletotrichum truncatum* (Schw.)	Brown blotch	All parts	[22] [37]
Fusarium oxysporum f.sp *tracheiphilum*	*Fusarium* wilt	Seedlings	[17]
Fusarium oxysporum f. sp *vasinfectum* (E. F. Smith) Synd & Hans	*Fusarium* wilt	Seedlings	[40]
Macrophomina phaseolina	*Macrophomina* blight	Seedlings (severe mortality)	[41]
Mycospharella cruenta Latham. (Anormoph of *Pseudocercospora*)	*Pseudocercospra* leaf spots	foliar	[3]
Phomopsis longicola	*Phomopsis* pod spot	Pods/fruits	[16]
Protomycopsis phaseoli Ramak & Subram. (=*Entyloma vinae* Batista)	Leaf smut	Foliar	[14]
Pythium aphanidermatum (Edison) Fitz	*Pythium* soft rot	Stem	[11]
Pythium ultimum	Damping off (pre/post)	Seed/seedling	[18]
Phytophthora cactorum (Leb. & Chon.) Schroet.	Red stem canker	Stem/root	[10]
Phytophthora vignae Puress.	*Phytophtora* stem rot	Stem/root	[10]
Sphaceloma sp.(Anamorph of *Elsinoe phaseoli* Jenkin)	*Sphaceloma* scab.	Hypocotyls & epicotyls	[3]
Sclerotiumrolfsii Sacc. (Teliomorph: *Corticum rolfsii* Curzi).	Basal stem rot/wilt	Stem	[42]
Uromyces appendiculatus (pers.) Unger (=*U. vinae* Barclay).	Brown rust	Foliar	[43]
Thanatephorus cumeris (Frank) Donk (=*Rhizoctonia solani* Kuhn)	Web blight Rot (root rot /seedling disease complex)	Root/seedling	[3]

Table 4. Plant families under the affliction of Colletotrichum Corda (1995-2011).

Family	Reference.
Amaranthaceae Juss.	[44]
Anacardiaceae Lindl.	[45]
Asteracea Bercht. & J. Presl.	[27] [44]
Brassicaceae Juss.	[46]
Carcaceae Dumort.	[30]
Convovulaceae Juss.	[27]
Cucurbitaceae Juss.	[44]
Cuscutaceae Dum. (=*Convovulaceae*)	[27]
Fabaceae Lindl.	[3] [8] [22] [44] [47]
Lauraceae Juss	[45]
Leguminosae Juss., Non. Con (=*Fabaceae* Lindl.)	[1] [3] [6] [9]
Linaceae. L	[37]
Malvaceae Juss	[44]
Musaceae Juss	[44]
Oleaceae Hoffmgg.& Link	[48] [49]
Poaceae Barnhart (=*Gramineae*Juss, Non. Con.)	[27] [50]
Roasaceae Adans	[44] [48] [51]
Rubiaceae Linn	[52] [53]

Table 5. Plant families screened for biofungicidal properties (1998-2011).

Family	Samples	Bioassay	Botanical Form	Reference
Alliaceae	Garlic spp	*In vitro*	Syrup	[38]
Annonaceae	*Xylopia aethiopica*; *Annona reticulata*	*In vitro/In vivo*	Oil extract; Aqueous.	[8] [54]
Arecaceae	*Elaeis guineensis*; *Cocos nucifera*	*In vitro*; *In vitro*	Ashes; Ashes	[34] [35]
Asteraceas	*Chromoleana odorata*; *Vernonia amygdalina*	*In vitro*; *In vivo*	Aqueous; Aqueous.	[25] [33]
Caricaceae	*Carica papaya*	*In vivo*	Aqueous.	[33]
Cochlospermaceae	*Cochlospermum planchonii*	*In vitro*	Aqueous.	[25]
Euphorbiaceae	*Ricinus comunis*; *Hymenocardia acida*; *Euphorbia prostrata*	*In vitro*; *In vitro*; *In vitro*	Aqueous; Aqueous; Ashes	[4] [25] [35]
Fabaceae	*Tephrosia vogelii*; *Senna alata*	*In vitro*; *In vitro*	Aqueous; Ashes	[2] [35]
Lamiaceae	*Thymus vulgaris*; *Ocimum gratissimum*; *O. sanctum*	*In vivo*; *In vivo*; *In vitro*	Oil extract; Aqueous; Aqueous/Syrup	[23] [25] [32] [55]
Lauraceae	*Cinnamon zeylanicum*	*In vivo*; *In vitro*	Oil extract ;Syrup	[38] [55]
Meliciaceae	*Azadiractha indica*	*In vitro/In vivo.*	Oil extract; Aqueous	[8] [25]
Myrtacea	*Psidium guajava.*	*In vitro*	Aqueous.	[25]
Piperaceae	*Piperaceae*	*In vitro/In vivo*	Aqueous/syrup; Syrup	[9] [38]
Plumbaginaceae	*Plumbago zeylanica*	*In vitro*	Ashes	[35]
Poaceae	*Cymbopogon citratus*	*In vitro/In vivo*; *In vitro*	Aqueous; Oil extract	[30] [32]
Potederiaceae	*Eichhomia crassipes*	*In vitro*	Ashes.	[34]
Rutaceae	*Citrus limon*	*In vitro*	Aqueous/Syrup; Aqueous.	[9] [25]
Solanaceae	*Nicotiana tabacum*	*In vitro*	Aqueous.	[4]

Table 6. Spatial interaction between screened and pathogen afflicted plant families.

Plant family screened for biofungicidal properties(A)[1]	Plant family under the affliction of *Colletotrichum*(B)	Existence of spacial family interaction between A & B = C	Percentage family interaction [C/(AorB) x 100]
Alliaceae	*Amaranthaceae* Juss.		
Annonacea	*Anacardiaceae* Lindl.		
Arecaceae	*Asteracea* Bercht. & J. Presl.		
Asteracea Bercht. & J.Presl.	*Brassicaceae* Juss.	*Asteracea* Bercht. & J. Presl.	
Caricaceae Dumort.	*Caricaceae* Dumort.	*Caricaceae* Dumort	
Cochlospermaceae	*Convovulaceae* Juss.		
Euphorbiaceae	*Cucurbitaceae* Juss.		
Fabaceae Lindl.	*Cuscutaceae* Dum. (=*Convovulaceae*)	*Fabaceae* Lindl.	
Lamiaceae	*Fabaceae* Lindl.		
Lauraceae Juss	*Lauraceae* Juss	*Lauraceae* Juss.	**27.78%**
Meliciaceae	*Leguminosae* Juss., Non. Con (=*Fabaceae* Lindl.)		
Myrtaceae	*Linaceae* L.		
Piperaceae	*Malvaceae* Juss		
Plumbaginaceae	*Musaceae* Juss		
Poaceae Barnhart	*Oleaceae* Hoffmgg. & Link	*Poaceae* Barnhart	
Potederiaceae	*Poaceae* Barnhart (=*Gramineae* Juss, Non. Con.)		
Rutaceae	*Roasaceae* Adans		
Solanaceae	*Rubiaceae* Linn		

[1]Nominal value of A or B equals to 18; Nominal value of C is equals to 5.

REFERENCES

1. Latunde-Dada, A.O., O'Connell, R.J., Nasa, C., Pring, R.J., Lucas, J.A. and Bailey, J.A. (1996) Infection Process and Identity of the Hemibiotrophic Anthracnose Fungus Colletotrichum destructivum from cowpea (Vigna unguiculata). Mycological Research, 100, 1133-1141.

2. Allen, D.J., Thottappilly, G., Emechebe, A.M. and Singh, B.B. (1998) Disease of Cowpea. In: Allen, D.J. and Lenne J.M., Eds., Pathology of Food and Pasture Legumes, CAB International, Wallingford, 267-324.

3. Emechebe, A.M. and Lagoke, S.T.O. (2002) Recent Advances in Research on Cowpea Diseases in Challenges and Opportunities for Enhancing Sustainable Cowpea Production. Proceeding of the World Cowpea Conference III, the International Institute of Tropical Agriculture (IITA), Ibadan, 4-8 September 2000, 94-123.

4. Akinbode, O.A. and Ikotun, T. (2008) Evaluation of Some Bioagents and Botanicals in In Vitro Control of Colletotrichum destructivum. African Journal of Biotechnology, 7, 868-872.

5. Owolade, O.F., Adediran, J.A., Akande, M.A. and Alabi, B.S. (2006) Effect of Application of Phosphorus Fertilizer on Brown Blotch Disease of Cowpea. African Journal of Biotechnology, 5, 343-347.

6. Adebitan, S.A. and Ikotun, T. (1996) Effect of Plant Spacing and Cropping Pattern on Anthracnose Colletotrichum lindemuthianum of Cowpea. Fitopatología Brasileira, 21, 5-12.

7. Adebanjo, A. and Bankole, S.A. (2004) Evaluation of Some Fungi and Bacteria for Control of Anthracnose Disease of Cowpea. Journal of Basic Microbiology, 44, 3-9.

8. Amadioha, A.C. and Obi, V.I. (1998) Fungitoxic Activities of Extracts from Azadiractha indica and Xylopia aethiopica on Colletotrichum lindemuthianum in Cowpea. Journal of Herbs, Spices, and Medicinal Plants, 6, 33-40.

9. Amadioha, A.C. (2003) Evaluation of some plant leaf extracts against Colletotrichum lindemuthianum in cowpea. Acta Phytopathologica et Entomologica Hungarica, 38, 259-265.

10. Fernando, W.G.D. and Linderman, R.G. (1997) The Effect of Mycorrhizal (Glomus intraradisces) Colonization on the Development of Root and Stem Rot (Phytophthora vignae) of Cowpea. Journal of the Natural Sciences, 25, 39-47.

11. Bankole, S.A. and Adebanjo, A. (1998) Efficacy of Some Fungal and Bacteria Isolates in Controlling Wet Rot Disease of Cowpea Caused by Pythium aphanidermatum. Journal of Plant Protection, 11, 37-43.

12. Smith, J.E., Christian, L. and Aveling, T.A.S. (1999) Infection Process of Colletotrichum dematium on Cowpea Stems. Mycological Research, 103, 230-234.

13. Amadi, J.E. (1995) Chemical Control of Cercospora Leaf Spot Disease of Cowpea (Vigna unguiculata (L.) Walp.). Agrosearch, 1, 101-107.

14. Santos, A.A., Quindere, M.A.W. and Melo, M.B. (1997) Evaluation of Cowpea Genetypes for Resistance to Cowpea Smut (Entyloma vignae). Fitopatologia Brasileira, 22, 77-78.

15. Munoz, M.A. and Tamayo, M.P.J. (1994) The Process of Choanephora cucurbitarum in Cowpea [Vigna unguiculata (L.) Walp.]. ASCOLF Informa, 20, 4-5.

16. Roy, K.W. and Ratnayake, S. (1997) First Report of Phomopsis lengicolla Infection of Cowpea Pods and Seeds in Mississippi. Plant Diseases, 81, 693.

17. Smith, S.N., Helms, D.M. and Temple, S.R. (1999) The Distribution of Fusarium Wilts of Blackeyed Cowpea within California Caused by Fusarium oxysporum f. sp tracheiphilum Race 4. Plant Diseases, 83, 694.

18. Aveling, T.A.S. and Adandonon, A. (2002) First Report of Pre and Post Emergence Damping-Off of Cowpea Caused by Pythium ultimum in South Africa. Plant Disease Reporter, 84, 922.

19. Schumann, L.G. and D`Arcy, C.J. (2006) Essential Plant Pathology. The American Phytological Society, 338.

20. Amusa, N.A., Ikotun, T. and Osikanlu, Y.O.K. (1994) Screening Cowpea and Soybean Cultivars for Resistance to Anthracnose and Brown Blotch

Diseases Using Phytotoxic Metabolites. African Crop Science Journal, 2, 221-224.

21. Fokung, C.N., Ikotun, T. Dixon, A.G.O. and Akem, C.N. (1997) First Report of Colletotrichum gloeosporioides f. sp. Manihot, Cause of Cassava Anthracnose Disease Being Seed Born and Seed Transmitted in Cassava. Plant Disease, 81, 695.

22. Latunde-Dada, A.O., O'Connell, R.J., Bowwyer, P. and Lucas, J.A. (1999) Cultivar Resistance to Anthracnose Disease of Cowpea (Vigna unguiculata (L.)Walp.) Caused by Colletotrichum destructivum O'Gara. European Journal of Plant Pathology, 105, 445-451.

23. Colpas, F.T., Schwan-Estrada, K.R.F., Stangarlin, J.R., Ferrarse, M.L., Scapim, C.A. and Bonaldo, S.M. (2009) Induction of Plant Defense Responses by Ocimum gratissimum L. Leaf Extracts. Summa Phytopathologica, 35, 191-195.

24. Jansch, S., Rombkeand, J. and Frische, T. (2009) Ecological Effects of the Accumulation of Copper in Soil. (Results of a Literature Review) Extended Abstract in Zum Gleichnamigen Vortragbeider, 4, 70.

25. Nduagu, C., Ekefanand, E.J. and wankiti, A.O.N. (2008) Effect of Some Crude Plant Extracts on Growth of Colletotrichum capsici (Synd) Butler and Bisby, Causal Agent of Pepper Anthracnose. Journal of Applied Biosciences, 6, 184- 190.

26. Polyversum (2012)

27. Amusa, N.A. and Ikotun, T. (1995) Phytotoxicity of Metabolites Produced by Some Colletotrichum Species on Weeds and Crop Plants in Nigeria. International Journal of Tropical Plant Diseases, 13, 113-119.

28. Bankole, S.A. and Adebanjo, A. (1996) Biocontrol of Brown Botch of Cowpea Caused by Colletotrichum truncatumwith Trichoderma viride. Crop Protection, 15, 633-636.

29. Adekunle, A.T., Cardwell, K.F., Floriniand, D.A. and Ikotun, T. (2001) Seed Treatment with Trichoderma Species for Control of Damping-Off of Cowpea Caused by Macrophomina phaseolina. Biocontrol Science and Technology, 11, 449-457.

30. Palhano, F.L., Vilches, T.T.B., Santos, R.B., Orllando, M.T.D., Ventura, J.A. and Fernandos, P.M.B. (2004) Inactivation of Colletotrichum gloeosporiodes Spores by High Hydrostatic Pressure Combined with Citral or Lemon Grass Essential Oil. International Journal of Food Microbiology, 95, 61-66.

31. Obi, V.I. and Onuoha, C.I. (2000) Extraction and Characterization Methods of Plants and Plant Products. In: Ogbulie, J.N. and Ojiako, O.A., Eds., Biological and Agricultural Techniques, Websmedia Publishers, Owerri, 271-286.

32. Amadioha, A.C. and Obi, V.I. (1999) Control of Anthracnose Disease of Cowpea by Cymbopogon citratus and Ocimum gratissimum. Acta Phytopathologica et Entomologica Hungarica, 34, 85-89.

33. Ogwulumba, S.I., Ugwuoke, K.I. and Iloba, C. (2008) Prophylactic Effect of Pawpaw Leaf and Bitter Leaf Extracts on the Incidence of Foliar Mycopathogens of Groundnut (Arachis hypogaea L.) in Ishiagu, Nigeria. African Journal of Biotechnology, 7, 2878-2880.

34. Enikumehinand, A.O. and Kehinde, I.A. (2007) In Vitro Screening of Some Tropical Ash Samples against Seed Borne Pathogen of Wheat (Tritticum aestivum L.). Australasian Plant Pathology, 36, 587-590.

35. Obi, V.I. and Ugwunze, N.N. (2009) Control of Tuber Roots of Colocasia esculenta L. Using Plant Ash. Recent Progress in Medicinal Plants, 23, 345-355.

36. Regulating Pesticides (2012).

37. Latunde-Dada, A.O. and Lucas, J.A. (2007) Localized Hemibiotrophy in Colletotrichum: Cytological and Molecular Taxonomic Similarities among C. destructivum, C. linicola and C. truncatum. Plant Pathology, 56, 437-447.

38. Win, N.K., Jitareerat, P., Kanlayanarat, S. and Sangchote, S. (2007) Effect of Cinnamon Extract, Chitosan Coating, Hot Water Treatment and Their Combination on Crown Rot Disease and Quality of Banana Fruit. Postharvest Biology and Technology, 45, 333-340.

39. Grangeand, N.L.A. and Aveling, T.A.S. (1998) First Report of Alternaria cassiae on Cowpea. Plant Disease, 82, 1171.

40. Ushamalini, C., Rajappan, K. and Gangadharan, K. (1997) Inhibition of Macrophomina phaseolina and Fusarium oxysporum f. sp. tracheiphilum by Antagonists under in Vitro Conditions. Plant Disease Research, 12, 168-170.

41. Ratnoo, R.S., Jain, K.L. and Bhatnagar, M.K. (1997) Effects of Atmospheric Temperature on the Development of Ashy Stem Blight of Cowpea. Journal of Mycology and Plant Pathology, 27, 90-91.

42. Muqit, A., Haque, M.S. and Hossain, M.M. (1996) Reaction of Cowpea Lines against Sclerotiumrolfii. Bangladish Journal of Plant Pathology, 12, 63.

43. Heath, M.C. (1998) Involvement Reactive Oxygen Species in the Response of Resistant (Hypersensitive) or Susceptible Cowpea to the Cowpea Rust Fungus. New Phytologist, 138, 251-263.

44. Liu, B., Wasilwa, L.A., Morelock, T.E., O'Neil, N.R. and Correll, J.C. (2007) Comparison of Colletotrichum orbiculare and Several Allied Colletotrichum s for mtDNA RFLPs, Intron RFLP and Sequence Variation, Vegitative Compatibility, and Host Specificity. The American Phytopathological Society, 97, 1305-1314.

45. Sanders, G.M., Korstenand, L. and Wehner, F.C. (2007) Survey of Fungicides Sensitivity in Colletotrichum gloeosporiodes from Different Avocado and Mango Production Areas in South Africa. European Journal of Plant Pathology, 106, 745-752.

46. Sun, H. and Zhang, J.-Z. (2009) Colletotrichum destructivum from Cowpea Infecting Arabidopsis thaliana (Wall Cress) and Its Identity to C. higginsianum. European Journal of Plant Pathology, 125, 459-469.

47. Mould, M.J.R., Robb, J. and Boland, G.J. (1992) Loss of Resistance to Colletotrichum trifolii in In Vitro Regenerated Alfalfa. Plant Cell, Tissue and Organ Culture, 28, 207-213.

48. Salazar, S.M., Castagnaro, A.P., Arias, M.E., Chalfoun, N., Tonelloand, U. and Diaz-Ricci, J.C. (2007) Induction of a Defense Response in Strawberry Mediated by a Virulent Strain of Colletotrichum. European Journal of Plant Pathology, 117, 109-122.

49. Gomes, S., Prieto, P., Martins-Lopes, P., Carvalho, T., Martin, A. and Guedes-Pinto, H. (2009) Development of Colletotrichum acutatum on

Tolerant and Susceptible Olea europaea L. Cultivars: A Microscopic Analysis. Mycopathologia, 168, 203-211.

50. Moore, J.W., Ditmore, M. and Tebeest, D.O. (2008) Pathotypes of Colletotrichum sublineolum in Arkansas. American Phytopathological Society, 92, 1415-1420.

51. Garrido, C., Carbu, M., Fernandez-Acero, F. J., Budge, G., Vallego, I., Colyerand, A. and Cantoral, J.M. (2008) Isolation and Pathogenicity of Colletotrichum s Causing Anthracnose of Strawberry in South West Spain. European Journal of Plant Pathology, 120, 409-415.

52. Moriwaki, J., Sato, T. and Tsukiboshi, T. (2003) Morphological and Molecular Characterization of Colletotrichum boninense sp. nov. from Japan. Mycoscience, 44, 47-53.

53. Nguyen, P., Petersson, O.V., Olsson, P. and Liljeroth, E. (2010) Identification of Colletotrichum Species Associated with Anthracnose Diseases of Coffee in Vietnam. European Journal Plant Pathology, 127, 73-87.

54. Bautista-Baños, S., Hernandez-Lopez, M., Bosquez-Molina, E. and Wilson, C.L. (2003) Effects of Chitosan and Plant Extracts on Growth of Colletotrichum gloeospiodes, Anthracnose Levels and Quality of Papaya Fruit. Crop Protection, 22, 1087-1092.

55. Montes-Belmont, R. and Carvajal, M. (1998) Control of *Aspergillus flavus* in Maize with Plant Essential Oils and Their Components. *Journal of Food Protection*, **61**, 616-619.

Chapter 6

Bioactive Plant Metabolites in the Management of Non-Communicable Metabolic Diseases: Looking at Opportunities beyond the Horizon

Chandan Prasad [1,2,*], Victorine Imrhan [1], Shanil Juma [1], Mindy Maziarz [3], Anand Prasad [4], Casey Tiernan [1] and Parakat Vijayagopal [1]

[1]Department of Nutrition and Food Sciences, Texas Woman's University, Denton, TX 76204, USA

[2]Department of Medicine, LSU Health Sciences Center, New Orleans, LA 70112, USA

[3]Department of Nutrition and Food Sciences, Texas Woman's University, Houston, TX 77030, USA

[4]Department of Medicine, Division of Cardiology, UT Health Science Center at San Antonio, 7703 Floyd Curl Drive, San Antonio, TX 78229, USA

ABSTRACT

There has been an unprecedented worldwide rise in non-communicable metabolic diseases (NCDs), particularly cardiovascular diseases (CVD) and diabetes. While modern pharmacotherapy has decreased the mortality in the existing population, it has failed to stem the rise. Furthermore, a large segment of the world population cannot afford expensive pharmacotherapy. Therefore, there is an urgent need for inexpensive preventive measures to control the rise in CVD and diabetes and associated co-morbidities. The purpose of this review is to explore the role of food bioactives in prevention of NCDs. To this end, we have critically analyzed the possible utility of three classes of food bioactives: (a) resistant starch, a metabolically resistant carbohydrate known to favorably modulate insulin secretion and glucose metabolism; (b) cyclo (His-Pro), a food-derived cyclic dipeptides; and (c) polyphenol-rich berries. Finally, we have also briefly outlined the strategies needed to prepare these food-bioactives for human use.

KEYWORDS

cardiovascular disease; diabetes; food bioactives; functional food; resistant starch; polyphenols; cyclo (His-Pro)

1. INTRODUCTION

We have made a remarkable progress in public health and healthcare, including managing communicable diseases in the last 100–150 years. Some of the notable human achievements include discovery of antibiotics, starting with first antibiotic by Flemming in 1930s [1], vaccines [2], clean water supplies and better sewage disposal and, finally, fortification of foods to combat micronutrient deficiencies in many parts of the world [3]. These have led to a significant increase in life expectancy throughout the world. For example, as per UN estimates life expectancy between 2000 and 2050 is projected to increase by 16 years even in some of the poorest nations of Africa [4]. This is the good news. However, something else has happened with the status of the health of world citizens that is perhaps more alarming. There is a steady upsurge in non-communicable metabolic diseases (NCDs), diabetes, cardiovascular disease (CVD), obesity, and others worldwide since the 1970s, if not earlier. For example, as per United Nations estimates, the incidence of diabetes and CVD in China and India, the two most populous nations of the world representing about one-third of the world population, is projected to double with a total number exceeding 100 million by year 2025 [5].

The reasons underlying this remarkable change in human health is complex and perhaps has a lot to do with our lifestyle than genetics. Therefore, it is imperative that we have a discussion on how our lifestyles have changed over the last decades. The period following WWII has seen relative peace, agricultural revolution in 1970s that brought significant reduction in world food shortages, and food fortification with micronutrients such as vitamins and minerals. While this brought better food security worldwide, there were other more recent changes that possibly may have impacted world lifestyle, hence health. Some of these include wide-spread imbalance in current dietary choices, overuse of refined hyper-processed foods, and low intake of fibers and plant-derived bioactives. All such factors are beyond the scope of this review. However, it would be appropriate to enumerate some of these factors that have been reviewed in detail elsewhere. Some of these include, global changes in income distribution [6], globalization leading to change in eating habits [7,8,9], rural migration and urbanization [10,11], and automobiles, television, and inactive entertainment instruments [12,13].

Based on extensive animal studies and human epidemiologic studies, we are certain that obesity plays a central role in the etiology of diabetes and CVD. Therefore, the major focus of this review will be to examine whether nutrition via delivery of functional foods or food bioactives could help reduce incidence of obesity, a precursor to two major NCDs—diabetes and CVD. Using management of dyslipidemia as an example, we have next summarized successes and failures of current use of statin therapy, as well as alternatives to statins.

2. SUCCESSES AND FAILURES OF CURRENT STRATEGIES FOR MANAGEMENT OF NON-COMMUNICABLE DISEASES: LIPID LOWERING THERAPY AS AN EXAMPLE

The use of HMG-COA reductase inhibitors (statins) has grown in frequency and intensity over the past two decades for secondary reduction of cardiovascular events and for the primary prevention of events in high risk individuals. The most recent iteration of the blood cholesterol guidelines put forth through the collaboration of the American College of Cardiology, the American Heart Association, and the National Heart Lung and Blood Institute have been met with controversy. These agencies have jointly suggested use of fixed dose, high- or moderate-intensity statin-based lipid-lowering strategies based upon the patient's risk profile rather than targeting specific low density lipoprotein (LDL) or high density lipoprotein (HDL) goals [14]. Evidence would suggest that the latest iteration of these guidelines is projected to increase the number of patients treated with statin therapy [15]. Along with this increased utilization will be the ever-present concern of "statin intolerance". This review will outline the therapeutic strategies based on both naturally occurring and man-made pharmaceutical agents to consider in statin intolerant patients.

2.1. Statin Intolerance

Statin intolerance is most commonly defined as the inability to continue statin therapy or tolerate increased intensity of therapy due to side effects with or without blood evidence of muscle or liver abnormalities. Although, the incidence of frank liver failure, or rhabdomyolysis, is exceedingly rare (less than 1 in 20 million), the development of milder symptoms of myalgias and fatigue are more common and is estimated to be present in 10%–18% of patients [16]. Muscular side effects remain a major source of statin intolerance and reason for discontinuation of therapy. Advanced age, female gender, low muscle mass, baseline liver disease, and concomitant use of gemfibrozil all increase the risk of developing statin intolerance [17,18]. Multiple different dosing strategies have been recommended to manage statin intolerance while maintaining therapy including: use of lower dosages, intermittent dosing, administration of slow release agents (fluvastatin), trial of alternative statins, and minimization of drug-drug interactions [18].

2.2. Coenzyme Q10

Although the complete etiology of the myalgias remains unclear, the reduction in coenzyme Q10 (CQ10) in myocytes has been implicated as a potential source for this adverse effect [19]. Empirically, the use of CQ10 supplementation has been used in patients with varying degrees of success. A meta-analysis of placebo-controlled trials of statin use confirmed a significant reduction in serum levels of CQ10 with statin therapy [20]. Unfortunately, there has not been consistent evidence from randomized data to support supplementation with CQ10 in patients with myopathy

[20]. A recent systematic review supports this clinical equipoise and, given the low risk of side effects of CQ10, suggests that individual patients may derive benefit—in part due to placebo effect [19].

2.3. Vitamin D Supplementation

Apart from CQ10, recent focus has turned to vitamin D levels as an important factor in the development of statin related myalgias. Using data from the National Health and Nutrition Examination Survey 2001–2004, Morioka *et al.*, demonstrated that middle aged and older adult statin users with serum 25-hydroxyvitamin D (25[OH]D) levels <15 ng/mL, had nearly two times greater odds of reporting musculoskeletal pain compared to non-statin users [21]. Whether this association is causative or not is unclear and at this time speculative hypotheses note that vitamin D deficiency can lead to atrophy of type II muscle fibers—the clinical manifestations of which could be magnified by statin use. A more recent prospective clinical study in statin intolerant patients with vitamin D levels <32 ng/mL, demonstrated that with sufficient vitamin D supplementation (50,000–100,000 units/week), up to 95% of patients could be maintained on statin therapy without symptoms or myonecrosis and this re-initiation of lipid-lowering therapy was associated with an improved cholesterol profile [22]. At this point, use of vitamin D remains attractive and may be an option in deficient patients, although large randomized studies and mechanistic data remain scarce.

2.4. Alternatives to Statins

In individuals who are truly intolerant to statin therapy, alternative agents may be considered. The common themes surrounding these therapies is a lower intensity of LDL lowering, relatively fewer cardiovascular outcomes data, and additional side effects that may complicate tolerability. The agents available in this context include vitamin B3 (niacin or nicotinic acid), fibrates (amphipathic carboxylic acids), cholestyramine (bile acid sequestrants), plant sterol esters, red rice yeast, eztimibe, and the newly approved proprotein convertase subtilisin/kexin type 9 (PCSK9) inhibitors. Although a full discussion of the data for each of these agents is beyond the scope of the present review, Table 1 summarizes the key clinical and pharmacological points.

2.5. PCSK9 Inhibition: The Future?

There is considerable interest in a newly approved class of medications, the PCSK9 inhibitors. These agents include evolocumab, alirocumab, and bococizumab, and are human monoclonal antibodies against PCSK9—administered as a subcutaneous injection every two weeks [23]. PCSK9 is a serine protease which binds to the LDL receptor leading to the intra-hepatocyte degradation of the receptor [24]. The effect of these drugs is to increase LDL receptor expression in the liver and facilitate LDL clearance—thereby lowering serum LDL cholesterol. While clearly an attractive treatment for patients with familial hypercholesterolemia who have mutations of the PCSK9 gene and are difficult to treat with statins, the potential of these agents lies in treatment of the larger population of dyslipidemic patients. The various agents appear to have a class effect in terms of LDL lowering with comparable magnitude of effect. The first agent to receive United States Food and Drug Administration (FDA) approval was alirocumab (July 2015) and, per the FDA, approval is indicated for additive therapy to dietary modification and maximally tolerated statin therapy in adult patients with heterozygous familial hypercholesterolemia or patients with clinical atherosclerotic cardiovascular disease such as heart attacks or strokes, requiring additional lowering of LDL cholesterol. Interestingly, dose escalation of statins up-regulates PCSK9 which explains the magnitude of LDL lowering. PCSK9 monotherapy has been shown to reduce LDL levels from 40% to over 50% [25]. When coupled with a statin (even at a low dose) the reductions range from 40% to 70%. The use of PCSK9 inhibitors in statin intolerant patients is particularly attractive given the relatively mild side effect profile. The GAUSS-2 (Goal Achievement after Utilizing an Anti-PCSK9 Antibody in Statin Intolerant Subjects) study examined the relative effect of LDL lowering by evolocumab compared to ezetimbide in patients who were statin-intolerant due to muscle-related side effects [26]. The LDL lowering of evolocumab in the study was 53%–56% as compared to 15%–18% with ezetimibe.

These data suggest a role of this new class of drug as monotherapy for LDL lowering in statin intolerant individuals. The missing link is in PCSK9 inhibition therapy is the impact of these drugs on cardiovascular outcomes. Nearly 60,000 high-risk patients are being evaluated in randomized controlled trials to address this question. In addition, long term safety data and elucidation of any pleiotropic effects remains to be determined.

Table 1. Agents commonly used for the management of dyslipidemia.

Agent/Dose	Mechanism of Action	Effects on Lipid Profile	Common Side Effects	Key Clinical Data
Niacin	Inhibits triglyceride (TG) synthesis and favorably impacts apoliprotein B containing lipoproteins	Two gram dose can lower low density liproprotein (LDL) by 15%, increase HDL by up to 25%, and lower TG by up to 30%	Flushing, gastrointestinal effects, pruritis, and rash	Randomized controlled trials of niacin including AIM-HIGH and HPS2-THRIVE have demonstrated no benefit in cardiovascular events despite significant increases in HDL levels [27,28].
Fibrates	Reduces hepatic TG production, enhances LDL uptake by the LDL receptor, and stimulates lipoprotein lipolysis	Primarily lowers TG and can lower LDL by ~10%	Gastrointestinal side effects including nausea and diarrhea. Can raise liver enzymes and cause gallstone formation. Drug-drug interactions with statins and warfarin are important to watch	Meta-analyses have demonstrated a reduction in coronary events including non-fatal MI. However, no benefit in all-cause mortality was seen [27,28]
Bile acid sequestrants	Bind to bile acids in the intestine and prevent recirculation of cholesterol	Depending on dosage, can decrease LDL 15%–30%	Gastrointestinal side effects including constipation, very rare reports of myalgias	May have beneficial impact on coronary atherosclerosis and may lower HbA1c [27].
Plant sterol esters	Interfere with dietary and biliary cholesterol absorption from the intestines	2–3 grams per day may reduce LDL cholesterol by 8%–15% (typically 8%–9%) and TG by 6%–9%	Well tolerated, uncertain impact on vitamin and nutrient absorption Uncertain impact on cancer risk	Limited data on markers of atherosclerosis and no large randomized cardiovascular outcomes data [29].
Red rice yeast	Functionally a low dose statin. Contains monacolin—the active ingredient in lovastatin is monacolin K	Reductions in LDL vary between 19% and 30% Minimal impact on HDL and modest effect on TG reduction	Generally well tolerated—even in statin intolerant patients but there have been reported myalgias similar to statins with one case of rhabdomyolysis reported.	Several small randomized controlled trials. Recent meta-analysis of 13 trials confirms LDL reduction, modest TG reductions and no impact on HDL levels [30]. No significant creatinine kinase level changes were noted.
Ezetimibe	Decreases cholesterol absorption in the small intestine	As an alternative to statin therapy can lower LDL cholesterol ~15%–20%. Minimal impact on HDL as monotherapy. With statin use can lower LDL by ~25%–60% with moderate reduction in TG. Can be used with fibrates to achieve ~20% lowering of LDL, and moderate increase in HDL, and decrease in TG	Generally well tolerated. Most common side effects: diarrhea (2.0%–2.5%), myalgias with statin use (3%), transaminitis with statin use (1.3%)	Clinical efficacy and safety for LDL lowering have been established in multiple studies. IMPROVE-IT randomized 18,144 recent acute coronary syndrome patients to simvastatin + ezetimibe *vs.* simvastatin + placebo [31]. There was a median follow up of six years and the primary composite endpoint was death from CVD, a major coronary event (nonfatal myocardial infarction (MI), documented unstable angina requiring hospital admission, or coronary revascularization occurring at least 30 days after randomization), or nonfatal stroke. The trial demonstrated an absolute risk reduction of 2.0% ($p = 0.016$) in the primary endpoint, driven by a significant reduction in non-fatal MI and stroke.

Based on over three decades of study, the LDL hypothesis as a central feature of atherosclerosis and adverse cardiovascular outcomes remains the focus of lipid management. Statin intolerance, though less common in randomized trials, is not uncommon in clinical practice. The majority of non-statin LDL-lowering drugs have modest LDL reductions and have additional side effects. The novel class of PCSK9 inhibitors appears to offer a new option for statin intolerant patients with profound reduction in serum LDL cholesterol without myalgias. The long term efficacy and safety of these drugs requires further evaluation. The next section presents a general overview food bioactives that may have a role to play in the management of CVD and diabetes.

3. A BRIEF SURVEY OF FOODS RICH IN BIOACTIVE PLANT METABOLITES

The relationship between diet and chronic diseases has been extensively studied. For many years research focused on reducing total fat, saturated fat, cholesterol, and trans-fat in the diet to reduce the risk of certain chronic diseases, such as CVD and stroke. However, dietary modifications that include plant products have been shown to reduce the risk of these chronic diseases. The Nurses' Health Study, the Health Professionals' Follow-up Study [32,33], Women's Health Study [34], and the Physicians' Health Study [35] show that increasing consumption of fruits and vegetables resulted in decreased risk for CVD. Similarly, Bazzano *et al.* [36] reported that consumption of fruits and vegetables was inversely associated with stroke incidence, stroke mortality, ischemic heart disease mortality, and CVD mortality. In addition to fruits and vegetables, grains have been reported to reduce the risk of CVD [37] and of all-cause mortality in postmenopausal women [38]. Known modifiable risk factors for CVD include smoking, sedentary lifestyle, diet, dyslipidemia, hypertension, obesity, and type 2 diabetes [39].

The observed protective effect of consuming plant foods on chronic diseases is likely due to their bioactive components. The plant bioactive compounds that will be discussed in this review include phytosterols, flavonoids, and phytoestrogens (lignans). We will focus on effects of these bioactive compounds on CVD and type 2 diabetes.

3.1. Phytosterols

Phytosterols are naturally-occurring plant sterols found in the non-saponifiable fraction of plant oils. Plants synthesize several types of phytosterols (e.g., sterols and stanols) that are structurally similar to cholesterol, except for the functional group substitutions on the sterol side chain at the C24 position. Both cholesterol [40] and phytosterols [41,42] are important components of cell membranes responsible for regulating membrane fluidity and permeability.

Of all the phytosterols that have been identified, beta-sitosterol (most abundant), campesterol, and stigmasterol comprise almost our entire intake of phytosterols. Since humans do not synthesize phytosterols, they must be obtained from the diet. The main dietary sources of naturally-occurring phytosterols are vegetable oils, nuts, grains and, to a lesser extent, fruits and vegetables [43]. Phytosterols intake has been reported to decrease blood cholesterol despite their poor absorption (<10%) from the intestine [44]. In the intestine, phytosterols displace cholesterol in mixed micelles and, thus, inhibit the absorption of cholesterol [45] reducing circulating LDL concentration [46,47], a known risk factor for CVD.

As a result of the reported health benefits of phytosterols, and the small amounts that occur naturally in foods, FDA in 2000 authorized the use of a health claim—"may reduce the risk of heart disease" for foods containing phytosterols (beta-sitosterol, campesterol, and stigmasterol). For such a claim to be made, the food must contain "at least 0.65 g per serving of plant sterol esters or 1.7 g per serving of plant stanol esters, and consumed twice a day with meals for a daily total intake of at least 1.3 g of sterol esters or at least 3.4 g of stanol esters, as a part of a diet low in saturated fat and cholesterol may decrease the risk of heart disease [48]. In response to health petitions received, the FDA in 2010 proposed new requirements for a claim to be made for fortification of foods with phytosterols. "Foods containing at least 0.5 g per serving of plant sterols/stanols (or plant sterols and stanols) eaten with meals or snacks for a daily total intake of 2 g as a part of a diet low in saturated fat and cholesterol, may reduce the risk of heart disease [49].

The food industry seized the opportunity to provide commonly consumed products that are fortified with phytosterols, such as Benecol™ and Take Control™. Benecol spread contains stanol esters derived from tall oil (pine tree wood pulp) and Take Control margarine contains sterol esters from soybeans. Consuming 2–3 g/d of phytosterols from these products resulted in approximately 14% reduction in LDL with no change in HDL [50,51,52]. Thus, both sterols and stanols are equally effective in lowering LDL concentration. In addition to margarines and spreads, a variety of food groups (grains, cereals, dairy products, and salad dressings) were fortified with various amounts of stanols and sterols. Doses as low as 0.8–1.0 gram/day resulted in approximately 5% reduction in LDL concentration with higher doses (2 g/d) having greater cholesterol-lowering effect (~10%) [53,54,55]. The cholesterol-lowering effect of phytosterols has been shown to last up to one year [56,57]. These observations have resulted in the recommendation from the National Cholesterol Education Program (NCEP) Adult Treatment Panel III to include two grams of plant sterol or stanol esters daily for optimal dietary therapy for elevated LDL [58]. Recently, guidelines for lowering LDL concentration issued by the American Heart Association include consumption of a diet that emphasizes vegetables, fruits, whole grains, legumes, and nuts [38], which are foods that contain naturally-occurring phytosterols.

3.2. Flavonoids

This group of polyphenols is the most abundant in the human diet. It can be classified into many different categories based on their structures. The most common flavonoids are flavones, flavanols, catechins, and anthocyanins, along with anthoxanthins [59,60]. Epidemiological studies show an inverse relationship between flavonoid intake and chronic diseases including CVD [61,62,63,64]. When compared to low (<19 mg/d) intakes of flavonoids, individuals with higher intake (30 mg/d) had approximately 50% reduction in CHD mortality [64]. Similar reductions were observed with the consumption of apples [62] or onions [61]. The protective effects of flavonoids can be linked to the antioxidant capacity of these compounds and their metabolites.

Red wines contain an abundance of polyphenols including phenolic acids (for example, gallic acid, and caffeic acid), stilbenes (resveratrol), and flavonoids (for example, catechin, epicatechin, quercetin, rutin) [65]. The most abundant polyphenol in red wine is caffeic acid, but gallic acid had the highest antioxidant activity [66]. The polyphenols found in red wine inhibit oxidation of LDL *in vitro* [67,68] and increase total antioxidant capacity in the serum [69]. The polyphenols in red wine have antithrombotic effects due to decreased platelet aggregation, reduced synthesis of pro-thrombotic and pro-inflammatory mediators, and decreased expression of adhesion molecules [70,71]. In addition, red wine polyphenols can induce vasorelaxation via nitric oxide synthesis [71]. The protective effect of red wine polyphenols is independent of the alcohol content. Freedman *et al.* [72] reported that purple grape juice (7 mL/kg body weight per day for 14 days) decreased platelet aggregation (58% *vs.* 39%), increased nitric oxide release, and suppressed superoxide production. According to Xiang *et al.* [73] resveratrol exhibited lower antioxidant activity than gallic, caffeic, and syringic acids. However, the addition of resveratrol to the wine samples did not significantly improve the antioxidant activity, suggesting that resveratrol may exert its health effects by other mechanisms [73].

Resveratrol is a polyphenol found principally in the skin of grapes and, in lesser amounts, in peanuts. Moderate consumption of red wine (French paradox) reduced the risk for CVD via several mechanisms. Unlike observation of Xiang *et al.* [73], others have shown resveratrol to effectively scavenge free radicals and other oxidants [74,75]. Resveratrol inhibits both LDL oxidation [63] and platelet aggregation [76], as well as cyclooxygenase-2 (COX-2) synthesis [77]. Platelet aggregation is one of the first steps in blood clot formation. Blood clots have been known to occlude arteries resulting in myocardial infarction or stroke. One of the earliest steps in the development of atherosclerosis is recruitment of inflammatory cells from the blood to the arterial wall by vascular cell adhesion molecules (VCAM), and *in vitro* studies show resveratrol decreased the expression VCAM-1 [78] and matrix metalloproteinase-9 (MMP-9) mRNA [79] by suppressing nuclear factor AP-1 activation [77]. Resveratrol may also inhibit pro-inflammatory transcription factors such as NFκB or AP-1 [80]. The cardioprotective effect seen in *in vitro* studies are supported by animal data. Rats pre-infused with resveratrol prevented reperfusion-induced arrhythmias and mortality [81]. Similar to cardioprotective effects, resveratrol improved insulin sensitivity, glucose tolerance, and lipid profiles in obese animals [82].

One study showed improved flow-mediated dilation (FMD) of the brachial artery when patients with metabolic syndrome consumed 100 mg/d of resveratrol for three months. However this improved blood flow lasted only three months after the study was completed [83]. In another study (84), overweight and obese adults were given a single dose of resveratrol (30 mg, 90 mg, or 270 mg). All three doses improved FMD at 60 min after consumption [84] and in a follow up study, FMD improved in subjects consuming a single dose of resveratrol (75 mg) or daily doses (75 mg) for six months [85]. Subjects consuming resveratrol had significantly reduced VCAM [86], reduced circulating inflammatory markers, C-reactive protein, tumor necrosis factor, and plasminogen activator inhibitor-1 [87] reduced oxidized LDL and apolipoprotein B [88].

Resveratrol consumption (1000 mg/d for 45 days) significantly lowered fasting glucose, insulin concentrations, hemoglobin A-1c levels, and insulin sensitivity (HOMA-IR) in subjects with type 2 diabetes [89]. Similar glycemic benefits were observed at a lower dose of 250 mg/d for three months [90], and at a very low dose of 10 mg/d for four weeks [91]. These data show that resveratrol consumption could improve cardiovascular health as well as type 2 diabetes control.

3.3. Lignans

Lignans are polyphenols found in plants, especially in flaxseed (secoisolariciresinol diglucoside), sesame seeds (sesamin, sesamolin), and soy, followed by whole-grains cereals (syringaresinol), and legumes, including nuts. Flaxseed contains approximately 335 mg lignans/100 g [92] and sesame seeds contain approximately 373 mg/100 g [93]. Fruits and vegetables contain a wide variety of lignans (e.g., matairesinol (MAT), pinoresinol (PINO) and lariciresinol (LARI)) but in minute quantities [94].

Consumption of lignans is associated with decreased risk of CVD. The proposed mechanisms by which dietary lignans could reduce the risk of CVD include the phytoestrogenic, and antioxidant activity of these compounds and their metabolites. First, some plant lignans such as matairesinol (MAT), secoisolariciresinol (SECO), pinoresinol (PINO), and lariciresinol (LARI) are metabolized by intestinal bacteria to enterolignans (enterodiol and enterolactone) in various proportions [95,96,97], absorbed, and transported into systemic circulation. Enterolignans, bind to estrogen receptors alpha and beta, which are found in the endothelium of blood vessels and exert weak estrogenic activity [98]. Additionally, they promote the synthesis of sex hormone-binding globulin and inhibit the activity of several enzymes including aromatase and 5a-reductase [99]. The second is associated with the antioxidant activity of lignans and their metabolites [100]. In middle age–elderly men and post-menopausal women living in Northern Italy SECO, LARI, and total lignan intake decreased plasma soluble intercellular adhesion molecule-1 (sICAM-1). Soluble intercellular adhesion molecule-1 is the circulating form of ICAM-1 that is expressed on cell surface of many cell lines including endothelial cells. Interaction between ICAM-1 and lymphocyte function-associated antigen LFA-1 increases leukocyte adhesion and migration across the endothelium. Both ICAM-1 and sICAM increase the risk of atherosclerosis. In the same study, MAT intake increased FMD, another measurement of the endothelial function of blood vessels providing further evidence that plants contain bioactive compounds which reduce the risk of CVD [101].

The next section describes in detail three unrelated groups of food bioactives with potentials as agents for the management of CVD and diabetes. These include resistant starch—a carbohydrate, and two classes of small molecules—one derived as a result of food processing/ manufacturing and the other endogenous to fruit berries.

4. SELECT EXAMPLES OF BIOACTIVE PLANT METABOLITES WITH POTENTIALS FOR MANAGEMENT OF NON-COMMUNICABLE DISEASES

4.1. Resistant Starch

Complex carbohydrates derived from starch contribute over half of humans' daily energy requirements. Starch is a homopolysaccharide made in plants and stored in granules. Amylose and amylopectin are two polymers found in starch and are identified based on the glycosidic bond linking the α-D-glucose monomers. Amylose is a linear polymer with α-(1,4) linkages while amylopectin has linear α-(1,4) linkages and α-(1,6) branch points.

Starch was once considered to be completely digestible because human enzymes could cleave the amylose and amylopectin linkages entirely in the small intestine. However, in 1982, Englyst discovered that some foods contain small amounts of starch that resist enzymatic hydrolysis [102]. This starch is known as resistant starch (RS) and escapes digestion in the small intestine entering the large intestine intact. Four types of RS (RS 1 through RS 4) were classified by Englyst ten years after its discovery, which are based on its physical, structural, or chemically-modified properties [103]. Dietary sources of RS 1 include partially milled grains and seeds. RS 2 can be found in raw potatoes, legumes, just-ripe bananas, and high-amylose maize (HAM). HAM is a functional ingredient containing up to 60% amylose produced by corn breeding technology and is utilized most often in research protocols. RS 3 results from retrograded foods, such as potatoes, cereals, and breads. Chemically- or physically-modified starch and resistant maltodextrins are known as RS 4 and 5, respectively. The average daily global consumption of RS ranges from 3 to 10 g [104]. RS is considered both a dietary and functional fiber, thus increasing consumption of RS could improve fiber intake and overall health [104,105].

The digestive properties of RS in the small and large intestine, or lack thereof, contribute to its bioactivity and subsequent metabolic outcomes. Due to lack of enzymatic hydrolysis, the direct contribution of glucose to blood from RS is minimal and allows for an attenuated post-prandial glycemic response. The acute blood glucose response can be specific to RS type, where RS4 can be more effective than RS2; however, consuming RS4 over 12 weeks did not significantly lower fasting blood glucose [106,107]. Although studies reporting lower glucose concentrations immediately following RS intake have similar amounts of available carbohydrate with the control, results are more pronounced when RS replaces a portion of fully digestible carbohydrate [108,109]. The reductions in blood glucose are also more apparent among individuals with metabolic syndrome or type 2 diabetes who consumed 40 g RS2 daily over a longer duration (8–12 weeks) [110,111]. Of more importance, peripheral insulin sensitivity (Si) also improved by approximately 20% in individuals with metabolic syndrome consuming the same amount or RS that elicited reductions in blood glucose [110,111]. Interestingly, a 33% improvement in peripheral Si resulted without concomitant changes in blood glucose after a meal tolerance test after healthy adults consumed 30 g RS2 over four weeks [112]. Improvements in Si have also resulted with lower amounts of RS2 intake. One study showed a 56.5% improvement in Si in men, but not women, after 15 g of RS2 were consumed daily for four weeks [113]. The women in this study had 26% higher Si at baseline than men, suggesting that improvements in Si may be more apparent among individuals who are insulin resistant. To date, only one long-term study examined Si in adults with well-controlled type 2 diabetes. In this study peripheral or hepatic SI did not improve but changes in blood glucose were observed after daily intake of 40 grams of RS2 for 12 weeks [114]. While RS2 intake improves glucose control and Si, especially among individuals with obesity-related metabolic disorders, the type, amount, and duration of RS consumption must be considered when examining the metabolic impact.

The production of short chain fatty acids (SCFA) from RS fermentation by gut microbiota in the large intestine further classifies it as a bioactive compound. The SCFA are capable of influencing risk, and even treatment, of NCDs such as diabetes and cancer through several mechanisms:

decreasing luminal pH, enhancing mineral absorption, and stimulating the release of two satiety peptides known as glucagon-like peptide -1 (GLP-1) and peptide tyrosine tyrosine (PYY) to the periphery [115,116]. GLP-1 is an incretin known to reduce blood glucose concentrations, control gut motility, and influence food intake [117]. The PYY isoform also influences appetite by binding to Y2 receptors in the hypothalamus, thus activating pro-opiomelanocortin neurons that correspond to increases energy expenditure [118]. Interestingly, several animal studies have observed significant increases in GLP-1 and PYY that correlate with reductions in total body adiposity and visceral fat after RS2 intake over time [119]. The mechanism of reduced adiposity may be related to enhanced fat oxidation and reduced lipogenesis [120,121] and increased uncoupling protein-1 in brown adipose tissue [122].

Unfortunately, the favorable changes in body composition and satiety peptides observed in animal models have not been replicated in human studies [123]. Several reasons why animals showed pronounced improvements in satiety peptides and body composition after RS intake than humans include the amount consumed, gut microbiota composition, differences in cecum size and physiology, and age at intervention [124]. RS can act as a prebiotic to selectively increase the concentration and viability of certain bacteria, such as *Ruminococcus bromii* [125]. However, several factors may influence the fermentative capabilities of microbiota, including diet and genetics [125]. In addition, it would be physically impossible for humans to consume the amount of RS administered to animals.

One human study reported enhanced post-prandial GLP-1 response in individuals with type 2 diabetes consuming 40 g RS2 daily for 12 weeks but concomitant changes in glucose or insulin metabolism or body composition were not observed [114,126]. Theoretically, elevated concentrations of GLP-1 should translate to decreased energy intake through the incretin effect and improvements in satiation. Intra-individual variation in gut microbiota may influence RS fermentation, the production of SCFA, and upregulation of GLP-1 [109]. Since the half-life of GLP-1 is 1–2 min, very high concentrations may be required to promote satiety [127]. RS metabolism is also highly variable ranging from 82% to 53% among healthy and hyperinsulinemic adults, respectively [128], suggesting some individuals may excrete a large portion of unmetabolized RS.

The risk of colon cancer and chronic kidney disease may also be influenced by dietary RS. The consumption of 40 g of butyrylated RS2 with red meat (300 g) decreased O6-methyl-2-deoxyguanosie adducts, improved SCFA excretion, and altered the microbiota profile after four weeks [129]. The attenuation of chronic kidney disease progression has also been demonstrated by improving creatinine clearance and reducing inflammation in rats after consuming RS2 [130].

In conclusion, RS is a bioactive, fermentable fiber that can assist with the prevention and improvement of NCDs including insulin resistance and cancer. More research is needed to determine the exact mechanism(s), amount, type, and duration of RS intake, and type of microbiota capable of fermenting RS, to produce optimal metabolic outcomes.

4.2. Cyclic Dipeptides

Cyclic dipeptides (also known as 2,5dioxopiperazines; 2,5-diketopiperazines; cyclo (dipeptides); or dipeptide anhydrides) are relatively simple compounds and, therefore, are among the most common peptide derivatives found in nature. Curtius and Gloebel synthesized the first cyclic dipeptide, cyclo (Gly-Gly), in 1888 [131]; however, their existence as a special group of compounds in nature was not recognized until early in the 20th century [132]. Between the late 1800s and early 1900s, many simple diketopiperazines such as cyclo (Gly-Gly) were synthesized for the sole purpose of examining their interesting physicochemical properties [132]. In later years, a variety of dipeptide diketopiperazines were shown to exist in protein and polypeptide hydrolysates as well as fermentation broths and cultures of yeast, lichens, and fungi [133]. Some of these diketopiperazines were thought to result from nonenzymatic cyclization of dipeptides and their amides, inasmuch as they are often formed during chemical and thermal manipulations, as well as during storage of peptides and proteins [134]. Consistent with a role for fermentation process in synthesis of cyclic dipeptides is the observation of high levels of cyclo (His-Pro) in foods that undergo fermentation and/or high heat treatment of protein-rich foods. Such examples are nutritional supplements (e.g., TwoCal HN and Jevity), milk, yogurt, sauces, and fermented fish [135,136,137].

Although cyclic dipeptides are ubiquitous in nature, only very few of these have so far been shown to possess any biological activity. These include, cyclo (His-Pro), cyclo (Leu-Gly), cyclo (Tyr-Arg), and cyclo (Asp-Pro). Of these only cyclo (his-Pro) has been shown to be endogenous to animal kingdom. The demonstration of cyclo (His-Pro) synthesis in mammals was an accidental observation. In 1976, while studying thyrotropin-releasing hormone (pGlu-His-ProNH_2, TRH) metabolism, we observed that *in vitro* incubation of [^{3}H-Pro]TRH with hamster hypothalamic extracts or intraventricular administration of [^{3}H-Pro]TRH into rat brain led to the formation of a new metabolite that was characterized to be cyclo (His-Pro). The pleiotropic nature of the biologic activities associated with its precursor TRH, and prevailing acceptance of the role of limited proteolysis in generating new biologically active peptides aroused interest in cyclo (His-Pro). Two major advances—development of specific radioimmunoassay for cyclo (His-Pro) [138] and commercial availability of large quantities of synthetic peptide—helped research in pharmacology of cyclo (His-Pro) and demonstration of its presence in biologic samples.

The research essentially progressed on two fronts—measurement and characterization of CHP in a variety of biological samples including food and evaluation of its potential biologic activities. The data presented in Table 2 summarizes data on measurement of CHP in a number of food products. To date CHP has been shown to exhibit a score of diverse biologic activities that have been reviewed in detail elsewhere [133,134,139,140,141]. Based on the prevailing data in the literature we feel that CHP may be a good potential candidate food bioactive for the management of insulin resistance. Therefore, for the purpose of this review we are going to focus on the role of CHP in insulin physiology in relation to diabetes. Since food intake is an important signal for peripheral excursions in circulating insulin and glucose, we will also review role of CHP in food intake where appropriate.

4.3. Cyclo (His-Pro), Food Intake, and Body Weight

Morley *et al.*, (1981) were the first to demonstrate that intracerebral administration of CHP to rats led to suppression of food intake induced by stress, fasting, or spontaneous feeding [142]. While many investigators [143,144,145] have supported this initial observation, Bowden *et al.*, (1988) could not demonstrate any anorectic role for CHP [146]. An anorectic or satiety-modulating role for CHP is supported by elevation of CHP in brains of hyperphagic obese Zucker rats [147] and fasted Sprague–Dawley rats [148] and changes in plasma CHP in patients (anorexia nervosa and bulimia) with satiety disturbances [149].

These initial observations supporting an appetite suppressant and satiety-inducing role of CHP led to further examination of its role in body weight control. Treatment of aged (16–18 months), obese, and insulin-resistant Sprague–Dawley rats with diet rich in CHP for eight weeks resulted into improvement of oral glucose tolerance and body weight control [150]. In a follow-up study, using genetically type 2 diabetic Goto–Kakizaki (G-K) rats and aged obese Sprague–Dawley (S-D) rats, these investigators made a similar observation following treatment with synthetic CHP and zinc combination [151].

4.4. Cyclo (His-Pro), Insulin Secretion, Glucose Metabolism, and Diabetes

The demonstration of appetite inhibitory function of CHP and changes in its level with feeding-fasting coupled with availability of specific radioimmunoassay of CHP, led us to explore its role in insulin physiology. In 1983, we demonstrated presence of CHP in rat pancreatic islet at a level 88-fold higher than whole pancreas [152]. Soon after this observation, Leduque *et al.*, (1987) using immunohistochemical techniques demonstrated localization of CHP within pancreatic alpha cells that also harbor glucagon/glicentin [153]. CHP may act here like an incretin in augmenting insulin secretion and, thus, reducing plasma glucose. Since CHP is localized in alpha and not beta-cells of the pancreas, it may act upon beta cells to stimulate insulin release via a paracrine mechanism. A further support for a possible role of CHP in glucose metabolism comes from a very rapid rise (maximum at 15 min) in blood CHP during oral glucose tolerance test (OGTT) in normal human subjects [154]. In support that CHP may act like an incretin of gut origin, Hilton *et al.*, (1990) observed an acute and reversible rise in plasma CHP after glucose ingestion in rats and the response was greater after oral glucose than after intravenous glucose [155].

Two lines of investigation support a possible role of CHP in insulin secretion and glucose metabolism. First, in a series of animal experiments, two groups of investigators independently demonstrated improvement in insulin sensitivity following consumption of raw vegetables [156] or spent brewer's yeast hydrolysate [157] diets both rich in CHP. Second, since plasma levels of CHP are altered by oral glucose ingestion [154], Prasad and co-workers wondered whether exogenous CHP might alter the insulin response to oral glucose ingestion. To this end, rats were given 3 g/kg oral glucose load with either saline or increasing doses of CHP and plasma levels of insulin, C-peptide, and glucose were measured. The results showed CHP to cause higher insulin excursions without any change in C-peptide suggesting that CHP may decrease hepatic insulin clearance. An examination of the relationships between the levels of CHP in obese women and parameters of insulin secretion (fasting C peptide/insulin molar ratio) further suggested a role for this peptide in HIC and hyperinsulinemia in obesity [158].

Table 2. Presence of cyclo (His-Pro) in some common foods. Nutritional supplements were products of Ross Laboratories, Columbus, OH (1–8), Norwich Eaton, Norwich, New Your (9), and Travenol Laboratories, Deerfield, IL (10 and 11). Protein source: [1]Casein + Soy, [2] Casein, [3] Essential amino acids, [4] Crystalline amino acids. ND = Not detected.

Food	Cyclo (His-Pro) Concentration	Reference
Noodle	18.8 ng/g	[136]
Potted Meat	40.9 ng/g	[136]
Nondairy creamer	30.0 ng/g	[136]
Hot Dog	18.1 ng/g	[136]
Ham	32.5 ng/g	[136]
Egg	5.7 ng/g	[136]
White Bread	21.8 ng/g	[136]
Tuna	510.4 ng/g	[136]
Fish Sauce	1291.8 ng/g	[136]
Dried Shrimp	1630.8 ng/g	[136]
Fresh Cow Milk	1.8 ng/g	[135]
Pasteurized Cow Milk	2.5 ng/g	[135]
Yogurt	4.2 ng/g	[135]
Spent Brewer's Yeast hydrolysate	674,000 ng/g	[157]
Nutritional Supplements		
(1) Ensure plus [1]	300 ng/mL	
(2) Ensure HN [1]	601 ng/mL	
(3) Ensure [1]	454 ng/mL	
(4) Pulmocare [2]	824 ng/mL	
(5) Enrich [1]	323 ng/mL	
(6) TwoCal HN [2]	4763 ng/mL	[137]
(7) Jevity [1]	4467 ng/mL	
(8) Osmolite [1]	123 ng/mL	
(9) Tolerex [3]	ND	
(10) Travasorb Hepatic [4]	ND	
(11) Travasorb Renal [4]	ND	

In Figure 1, we have made a graphic attempt to visualize the complex role of CHP in controlling insulin secretion and glucose both in short and long terms at the level of pancreatic alpha and beta cells and liver acting as incretin, as well as autocrine, paracrine, and endocrine mediators. While our assumption of CHP stimulation of glucagon secretion remains to be examined, human beta cells express functional glucagon receptors which can, similar to incretin hormone receptors, generate synergistic signals for glucose-induced insulin secretion [159,160,161].

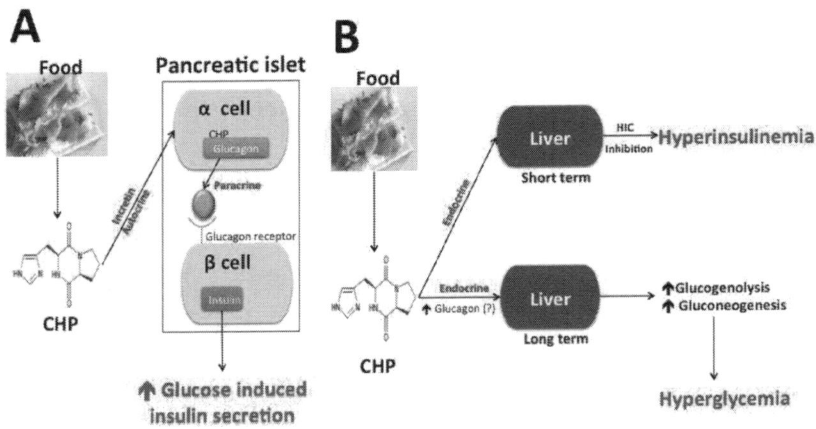

Figure 1. A diagrammatic representation of actions of Cyclo (His-Pro) at the level of the pancreas and liver in controlling insulin secretion and glucose metabolism. CHP = Cyclo (His-Pro). Panel A refers to action of CHP on pancreatic α and β cells via autocrine/incretin mechanisms. Panel B refers to short-term and long-term actions of CHP on liver.

While the mechanism underlying changes in hepatic insulin clearance is not clearly deciphered, prolonged decrease in HIC has been clearly associated with risks for insulin resistance, dyslipidemia, metabolic syndrome, and CVDs [162,163,164]. Acutely, CHP may help glucose metabolism by enhancing glucose-induced insulin secretion, a benefit to those with insulin resistance.

This opens further possibilities of new lines of investigation into the role of CHP and similar bioactive food components in the etiology of metabolic diseases.

4.5. Fruit Berries

Berries and other fruits rich in polyphenols and other natural compounds with bioactive properties have shown promise in recent years for use in the prevention of NCDs such as arthritis, diabetes, cancer, and CVD [165,166,167,168,169]. Antioxidant vitamins found in berries, such as vitamins C and E, may be important contributors to these outcomes [168]. Polyphenols found in berries and other plant foods are particularly associated with anti-inflammatory, antioxidant, cardioprotective, and chemopreventive properties, making them an ideal subject of research for chronic disease applications [167,168,169]. The mechanism of action is not fully understood, but appears to include down-regulation of the inflammatory cytokines, or by antioxidant or anti-inflammatory pathway signaling [166].

4.6. Bioactive Berry Constituents

A wide variety of phenolic compounds act as a protective mechanism against injury in the fruit. These compounds contribute to the antioxidant properties of berries and are typically found in the outer parts of the fruit or berry, most often as cinnamic and/or benzoic acid derivatives [166]. Anthocyanins tend to be located in the skin of the fruit, but are less strongly correlated to antioxidant activity than total phenolic content. Tannins contribute to sensory properties, like the tart taste of berries, and act as a stabilizing factor for the anthocyanins present. Small amounts of carotenoids and stilbenes such as resveratrol are present as well. While a well-known stilbene is the resveratrol found in grapes, smaller amounts of resveratrol can be found in cranberries, strawberries, and other berries. The antioxidant activity of total phenolic compounds can vary, however their Trolox Equivalent Antioxidant Capacity (TEAC) and Trolox values have been shown to be three to 30 times that of ascorbic acid and reference standards [166]. There is likely a synergistic activity among the differing bioconstituents present, with polyphenolic compounds accountable for most of the bioactivity [167].

Chokeberry, bilberry, and blackcurrant berries have the highest antioxidant capacity of the different berry fruits (umol Trolox/g fresh weight), and whole fruit extracts have greater antioxidant activity than many isolated phenolic compounds or vitamins [170]. However, more common fruits also contain bioactive compounds that may reduce the risk of chronic disease.

Strawberries are known to be high in phenolic compounds such as the phenolic acid derivative ellagic acid, and contain a significant amount of vitamin C. Blueberries are noted for a wide variety of anthocyanin compounds, while both cranberries and blueberries also contain significant concentrations of phenolic acids. The content and antioxidant activity of berry bioactive constituents can be dependent upon plant varietals, and growing, harvest, and storage conditions, as well as processing methods [170]. Since the bioavailability of these compounds differs between individuals [168], there is also a likely genetic component to the effectiveness of berry consumption in preventing NCDs.

4.7. Berry Consumption and Non-Communicable Disease

Of all the NCDs, most research regarding berry intake has been conducted in the area of CVD [165,171,172,173,174,175,176,177]. Berries may positively impact a number of preventable risk factors for CVD, including undesirable lipid profiles and hypertension [165,171,173,174, 175,176, 177,178]. This may be explained via a wide variety of mechanisms, including anti-inflammatory activity, free radical scavenging and up-regulation of antioxidant enzyme genes, decreased levels and antioxidation of LDL, increases in circulating HDL, inhibition of platelet activation and aggregation, and improvements in endothelial function [165,172,176, 179,180]. Other observed effects include decreases in hepatic cholesterol formation and fatty acid synthesis, and increases in cholesterol excretion and mitochondrial fatty acid oxidation [181]. Therefore, incorporation of berries and berry products into the diet may represent an accessible and affordable way to decrease cardiovascular risk factors.

Improved plasma antioxidant activity may be due in part to an increased vitamin C intake after regular berry consumption [167]. Del Bo and colleagues detected improvements in oxidative status following a 300 g dose of blueberries, but these were not accompanied by significant changes in peripheral arterial function or NO levels in a group of healthy males [182]. In contrast, a number of studies found increases in antioxidant activity were accompanied by desirable changes in lipid profiles [165,167,181]. Hypercholesterolemic male Dankin Hartley guinea pigs on a blueberry-supplemented diet for 75 days responded with reductions in oxidative stress, serum total cholesterol (TC) and LDL, and hepatic and aortic cholesterol accumulation. However, these observations were only in those animals given a high cholesterol diet. In this high cholesterol group, daily blueberry treatment equal to approximately 3.8–4 g fresh blueberries also decreased serum amino transferases, malondialdehyde (MDA), and diene conjugate (DC), and increased hepatic glutathione transferase (GST) activity when compared to the high cholesterol diet without blueberries [181].

4.8. Cardiovascular Diseases

Other promising studies have linked berry consumption to antioxidant activity with accompanying lipid changes in humans [165,167,171,177]. In healthy adult men, two weeks of 7 mL/kg cranberry juice intake resulted in increased antioxidant capacity from baseline (+6.5% ± −10.3, $p = 0.014$) and reduced oxidized LDL (−9.9% ± −17.8, $p = 0.0131$) [177]. In another study, 12 weeks of powdered strawberry beverage consumption was followed by a dose-responsive decrease in serum MDA (a marker of lipid peroxidation) in a group of abdominally obese adults with high serum lipids (low dose: 1.3 ± 0.2 umol/L, high dose: 1.2 ± 0.1 umol/L *vs.* low control: 2.1 ± 0.2 umol/L and high control: 2.3 ± 0.2 umol/L, $p < 0.05$). The low (25 g/d) and high doses (50 g/d) were approximately equal to 250 g and 500 g of fresh strawberries and compared to a control diet matched in fiber and calories. There were also dose-responsive decreases in serum TC, LDL cholesterol, and nuclear magnetic resonance (NMR)-derived small LDL particle concentration, which is associated with atherogenicity. However, adiposity, blood pressure, glycemia and serum HDL, triglycerides, C-reactive protein, and adhesion molecules did not significantly change [165]. In a third study, consumption of 500 g fresh strawberries for one month increased ferric-reducing antioxidant power and oxygen radical absorbance capacity measures in healthy adults (+24.97%, +41.18%, +41.36%, respectively, $p < 0.05$). Serum MDA, urinary 8-OHdG and isoprostanes decreased (31.40%, 29.67%, 27.90%, respectively, $p < 0.05$), as did spontaneous hemolysis, oxidative hemolysis, and activated platelets ($n = 10$, $p < 0.05$). As red blood cell hemolysis can be influenced by membrane lipid peroxidation, these results suggest an antioxidative effect of strawberry consumption. These changes were accompanied by further significant reductions in TC, LDL and triglycerides compared to baseline, (−8.78%, −13.72% and −20.80%, respectively, $p < 0.05$). It is possible the 2 g fiber/100 g fruit (approximately 10 g fiber daily above a normal diet) contributed to the observed decreases in cholesterol in this study [171].

4.9. Hypertension

Another benefit of berry consumption prominent in the available literature has been improvements in blood pressure or hypertensive status [183]. In normotensive male Wistar rats, ten weeks of supplementation with 2% (*wt/wt*) freeze-dried blueberry resulted in decreases in systolic blood pressure of 11%–14% compared to the control group. The supplement dosage correlates to approximately 450 g daily of freeze-dried blueberry intake in humans. Improvements were also seen in vascular reactivity, but not mean body weight or lipid profile [172]. Hypertension in rats is linked to impaired NO-dependent endothelial function, therefore the observed effects may be due to increased NO bioavailability via activation of endothelial NO synthase. Circulating polyphenol metabolites have also been shown to inhibit NADPH oxidase, which would improve the bioavailability of NO and, thus, also improve endothelial-dependent vasodilation [172].

Desirable changes in blood pressure have been also demonstrated after berry consumption in adults with cardiovascular risk factors [167,175,176,184]. In a group of 72 older subjects (n = 46 women, mean age 58) with one or more risk factors, an average daily intake of 837 mg mixed berry products resulted in modestly decreased systolic blood pressure (SBP) compared to an increase in the control group (−1.5 mm Hg *vs.* +0.5 mm Hg). While diastolic blood (DBP) pressure remained unchanged with treatment, it increased somewhat in the control group (+0.9 mm Hg). A subgroup of participants with baseline high blood pressure demonstrated even greater effects, with a mean decrease of 7.3 mm Hg in the treatment group compared to 0.22 mm Hg in the control ($p = 0.024$). Serum HDL also increased slightly in the berry group compared to control group (5.2% *vs.* 0.6%) [175]. In post-menopausal women with pre- and stage-one hypertension (n = 24 per group), daily consumption of 22 g freeze-dried blueberry powder (equal to about one cup fresh blueberries) resulted in decreased SBP and DBP compared to the control group (131 ± 17 mm Hg, $p < 0.05$ and 75 ± 9 mm Hg, $p < 0.01$, respectively). After eight weeks mean SBP was still considered pre-hypertensive, despite the decrease. Brachial-ankle pulse wave velocity, the gold standard marker for arterial damage and predictive of CVD risk, was also lower at the final visit with blueberry treatment (1401 ± 122 cm/s, $p < 0.01$). In addition, NO levels in the blueberry group were higher compared to baseline (9.11 ± 7.95 umol/L), but this was not the case in the control group [184]. In sedentary men and women without cardiovascular risk factors, consuming 250 g freeze-dried blueberry powder daily resulted in decreases in augmentation index (AIx) and aortic systolic pressure (ASP; $p = 0.24$ and $p = 0.46$, respectively) after six weeks of treatment. In addition, a subset of pre-hypertensive patients (n = 9) demonstrated significant reductions in DBP ($p = 0.38$) [185]. ASP is linked to increased CVD risk, and decreased AIx may be related to the decrease in aortic pressure. Interestingly, in a separate shorter study using the same dose in smokers there were no changes in blood pressure [186]. These results suggest that some cardiovascular risk factors, such as elevated blood pressure could be attenuated with berry intake, perhaps via increased NO production which mediates vasodilation.

4.10. Diabetes Mellitus

Other studies have demonstrated cardiovascular benefits after berry consumption in adults with diabetes or the metabolic syndrome [174,187,188] *In vitro* studies suggest that anthocyanins may influence gene expression in adipose tissue, which could partially account for a mechanism [189,190]. Six weeks of a 50 g daily strawberry beverage supplement in middle-aged adults with type 2 diabetes (mean age 51.57 ± 10 years) resulted in significantly decreased TC and TC:HDL ratio compared to baseline, but not compared to control (−13.8% and −7.1%, respectively, $p < 0.00$). SBP was also decreased in both treatment groups, but no significant difference existed between groups. However, the strawberry group did have significantly reduced DBP compared to placebo (78.7 ± 7.2 *vs.* 84.4 ± 5.8, $p = 0.01$), and within the strawberry group (84.2 ± 8.03 to 78.7 ± 7.2, $p = 0.00$) [188]. Six weeks of treatment with a twice-daily blueberry smoothie (total 45 g/d, equal to about 2 cups of fresh blueberries) in middle aged adults with the metabolic syndrome resulted in altered endothelial function despite having no effect on blood pressure or insulin sensitivity. Mean resting endothelial function (reactive hyperemia index or RHI) was significantly improved compared to the control group ($p = 0.024$), which persisted after adjustment for confounding factors ($p = 0.0023$). While not significant, 73% of the blueberry group experienced improved endothelial function and 61% of the placebo group experienced decreased endothelial function. It is possible that the absence of change in blood pressure may have been related to antihypertensive medications used by participants, or the more thorough 24-hour ambulatory pressure monitoring. In contrast, no significant changes in peripheral arterial function or NO levels were observed after blueberry intake in adults with the metabolic syndrome [187]. However in an earlier study these researchers did observe improved insulin sensitivity in obese adults [191], an effect also seen in mice fed a high-fat diet [192]. Another study in middle-aged obese men and women with metabolic syndrome used a similar (50 g) dose of freeze-dried blueberry powder, which resulted in SBP lowered by 6% and DBP lowered by 4% compared to the control group (−1.5%, −1.2%) [174].

While the initial literature on the benefits of berry consumption is promising, it is important consider berry types that have established availability and affordability for common use as fresh produce is seasonal and often expensive. However, if consumption of fruits such as berries can be definitively linked to prevention of non-communicable disease, it may be a bargain compared to the cost and side effects associated with conventional treatment. Berries are widely available in fresh or processed forms, but their polyphenolic constituents are more bioavailable in their naturally-occurring form than in supplemental forms [191,192]. Currently the available literature does not allow for definitive recommendations for berry type, dosage or form for a given disease state. Future studies should investigate biomarkers of polyphenol metabolites, and continue to study the effects of whole berry intake in humans with chronic diseases. Despite these limitations, berries are poised to become a prime functional food in the fight against NCDs.

5. FUTURE DIRECTIONS

The use of pharmacologic means for the management of NCDs, such as CVD and diabetes, has played a monumental role in decreasing mortality rate if not morbidity. These pharmacologic agents are generally designed to mimic receptor agonist/antagonist, enzyme inhibitors/stimulators, or supply missing substrates. Since most receptors or enzymes are ubiquitous to many organs and cell types, it is not surprising that these agents bring with them a plethora of side effects. We must emphasize here that use of pharmacologic agents in disease management is of utmost importance. However, considering side effects of treatment, there is an urgent need of search of preventative measures that may delay or abrogate use of pharmacological agents altogether.

Food bioactives are steadily appearing to be in a position to play this preventive role. However, there remain many technological issues in the way of bringing bioactives from bench to bedside. Below is a list of four major issues associated with the development of food bioactives for disease prevention. These include, (i) what is the nature of food bioactives in functional foods; (ii) how to establish limit of maximum purity without losing functionality of food bioactives; (iii) how to establish limit of maximum purity without losing shelf life of the food bioactives; and (iv) how to deliver efficiently and maximize half-life *in vivo*.

Food bioactives include a large class of molecules such as polyphenols, non-fermentable carbohydrates, w-3 fatty acid, beta-carotene, and sulfur-rich compounds, to name a few. To date >5000 phytochemicals have been identified as polyphenols while many more still remain unknown [193] and identified before their health benefits are fully understood. For example, risk of cancer is inversely related to the consumption of green and yellow vegetables and fruit. Since β-carotene is present abundantly in these vegetables and fruit, role of pure β-carotene in cancer prevention has been investigated extensively. However, the role of carotenoids as anticancer supplements has been questioned as a result of several clinical studies [194]. Such observations are not limited to β-carotene only. The reason for such disappointing results may be many. For example, to elicit its anti-cancer properties, β-carotene may need a congener or another food bioactive also endogenous to fruit and vegetables. Fruits and vegetables may have chemical components much needed to maintain bioavailability of β-carotene.

A better understanding of the nature of food bioactives, and their combinations and mechanism of their action in eliciting bioactivity, preservation of their bioactivity in storage, and a better understanding of how to increase their bioavailabilty is needed before effective utilization of food bioactives for disease prevention is achieved. To help bring food bioactives at the center stage of prevention, research must be a cooperative effort between all with expertise in nutrition, food science, analytical chemistry, separation science, pharmacology, and nanotechnology.

ACKNOWLEDGMENTS

This research was partly funded by the Human Nutrition Research Fund, State of Texas. The Freeman Heart Association Endowment in Cardiovascular Diseases supported AP as endowed chair.

AUTHOR CONTRIBUTIONS

All authors participated equally in preparation of this manuscript.

REFERENCES

1. Aldridge, S.; Parascadola, J.; Sturchio, J.L.; American Chemical Society; Royal Society of Chemistry (Great Britain). *The Discovery and Development of Penicillin 1928-1945: The Alexander Fleming Laboratory Museum, London, UK, November 19, 1999: An International Historic Chemical Landmark*; American Chemical Society: Washington, DC, USA, 1999.

2. Stern, M.N.; Markel, H. The history of vaccines and immunization: Familiar patterns, new challenges. *Health Aff.* **2005**,*24*, 611–621.

3. Bishai, D.; Nalubola, R. The history of food fortification in the United States: Its relevance for current fortification efforts in developing countries. *Econ. Dev. Cult. Chang.* **2002**, *51*, 37–53.

4. World Health Organization (WHO). Projections of Mortality and Burden of Disease, 2004–2030. Available online: http://www.who.int /healthinfo/global_burden_disease/projections2004/en/ (accesses on 13 September 2015).

5. King, H.; Aubert, R.E.; Herman, W.H. Global Burden of Diabetes, 1995–2025 Prevalence, numerical estimates and projections. *Diabetes Care* **1998**, *21*, 1414–1431.

6. Di Cesare, M.; Khang, Y.-H.; Asaria, P.; Blakely, T.; Cowan, M.J.; Farzadfar, F.; Guerrero, R.; Ikeda, N.; Kyobutungi, C.; Msyamboza, K.P.; *et al.* Inequalities in non-communicable diseases and effective responses. *Lancet* **2013**, *381*, 585–597.

7. Mintz, S.W.; Du Bois, C.M. The Anthropology of Food and Eating. *Annu. Rev. Anthropol.* **2002**, *31*, 99–119.

8. Hu, F.B. Globalization of food patterns and cardiovascular disease risk. *Circulation* **2008**, *118*, 1913–1914.

9. Hu, F.B. Globalization of Diabetes. The role of diet, lifestyle, and genes. *Diabetes Care* **2011**, *34*, 1249–1257.

10. Torun, B.; Stein, A.D.; Schroeder, D.; Grajeda, R.; Conlisk, A.; Rodriguez, M.; Mendez, H.; Martorell, R. Rural-to-urban migration and cardiovascular disease risk factors in young Guatemalan adults. *Int. J. Epidemiol.* **2002**, *31*, 218–226.

11. Holmboe-Ottesen, G.; Margareta Wandel, M. Changes in dietary habits after migration and consequences for health: A focus on South Asians in Europe. *Food Nutr. Res.* **2012**.

12. Hancox, R.J.; Milne, B.J.; Poulton, R. Association between child and adolescent television viewing and adult health: A longitudinal birth cohort study. *Lancet* **2004**, *364*, 257–262.

13. Douglas, M.J.; Watkins, S.J.; Gorman, D.R.; Higgins, M. Are cars the new tobacco? *J. Public Health* **2011**, *33*, 160–169.

14. Stone, N.J.; Robinson, J.G.; Lichtenstein, A.H.; Bairey Merz, C.N.; Blum, C.B.; Eckel, R.H.; Goldberg, A.C.; Gordon, D.; Levy, D.; Lloyd-Jones, D.M.; *et al.* 2013 ACC/AHA guideline on the treatment of blood cholesterol to reduce atherosclerotic cardiovascular risk in adults: A report of the American College of Cardiology/American Heart Association Task Force on Practice Guidelines. *J. Am. Coll. Cardiol.* **2014**, *63*, 2889–2934.

15. Tran, J.N.; Caglar, T.; Stockl, K.M.; Lew, H.C.; Solow, B.K.; Chan, P.S. Impact of the New ACC/AHA Guidelines on the Treatment of High Blood Cholesterol in a Managed Care Setting. *Am. Health Drug Benefits* **2014**, *7*, 430–443.

16. Bruckert, E.; Hayem, G.; Dejager, S.; Yau, C.; Begaud, B. Mild to moderate muscular symptoms with high-dosage statin therapy in hyperlipidemic patients—The PRIMO study. *Cardiovasc. Drugs Ther.* **2005**, *19*, 403–414.

17. Wilmot, K.A.; Khan, A.; Krishnan, S.; Eapen, D.J.; Sperling, L. Statins in the elderly: A patient-focused approach. *Clin. Cardiol.* **2015**, *38*, 56–61.

18. Vandenberg, B.F.; Robinson, J. Management of the patient with statin intolerance. *Curr. Atheroscler. Rep.* **2010**, *12*, 48–57.

19. Marcoff, L.; Thompson, P.D. The role of coenzyme Q10 in statin-associated myopathy: A systematic review. *J. Am. Coll. Cardiol.* **2007**, *49*, 2231–2237.

20. Littlefield, N.; Beckstrand, R.L.; Luthy, K.E. Statins' effect on plasma levels of Coenzyme Q10 and improvement in myopathy with supplementation. *J. Am. Assoc. Nurs. Pract.* **2014**, *26*, 85–90.

21. Morioka, T.Y.; Lee, A.J.; Bertisch, S.; Buettne, R.C. Vitamin D status modifies the association between statin use and musculoskeletal pain: A population based study. *Atherosclerosis* **2015**, *238*, 77–82.

22. Khayznikov, M.; Hemachrandra, K.; Pandit, R.; Kumar, A.; Wang, P.; Glueck, C.J. Statin Intolerance Because of Myalgia, Myositis, Myopathy, or Myonecrosis Can in Most Cases be Safely Resolved by Vitamin D Supplementation.*N. Am. J. Med. Sci.* **2015**, *7*, 86–93.

23. Shimada, Y.J.; Cannon, C.P. PCSK9 (Proprotein convertase subtilisin/kexin type 9) inhibitors: Past, present, and the future. *Eur. Heart J.* **2015**, *36*, 2415–2424.

24. Lambert, G.; Sjouke, B.; Choque, B.; Kastelein, J.J.; Hovingh, G.K. The PCSK9 decade. *J. Lipid Res.* **2012**, *53*, 2515–2524.

25. Koren, M.J.; Scott, R.; Kim, J.B.; Knusel, B.; Liu, T.; Lei, L.; Bolognese, M.; Wasserman, S.M. Efficacy, safety, and tolerability of a monoclonal antibody to proprotein convertase subtilisin/kexin type 9 as monotherapy in patients with hypercholesterolaemia (MENDEL): A randomised, double-blind, placebo-controlled, phase 2 study. *Lancet* **2012**, *380*, 1995–2006.

26. Stroes, E.; Colquhoun, D.; Sullivan, D.; Civeira, F.; Rosenson, R.S.; Watts, G.F.; Bruckert, E.; Cho, L.; Dent, R.; Knusel, B.; *et al.* Anti-PCSK9 antibody effectively lowers cholesterol in patients with statin intolerance: The GAUSS-2 randomized, placebo-controlled phase 3 clinical trial of evolocumab. *J. Am. Cardiol.* **2014**, *63*, 2541–2548.

27. De Groot, L.J.; Beck-Peccoz, P.; Chrousos, G.; Dungan, K.; Grossman, A.; Hershman, J.M.; Koch, C.; McLachlan, R.; New, M.; Rebar, R.; *et al.* *Endotext*; MDText.com, Inc.: South Dartmouth, MA, USA, 2000.

28. Keene, D.; Price, C.; Shun-Shin, M.J.; Francis, D.P. Effect on cardiovascular risk of high density lipoprotein targeted drug treatments

niacin, fibrates, and CETP inhibitors: Meta-analysis of randomised controlled trials including 117,411 patients. *BMJ* **2014**.

29. Gylling, H.; Simonen, P. Phytosterols, Phytostanols, and Lipoprotein Metabolism. *Nutrients* **2015**, *7*, 7965–7977.

30. Li, Y.; Jiang, L.; Jia, Z.; Xin, W.; Yang, S.; Yang, Q.; Wang, L. A meta-analysis of red yeast rice: An effective and relatively safe alternative approach for dyslipidemia. *PLoS ONE* **2014**, *9*, e98611.

31. Cannon, C.P.; Blazing, M.A.; Giugliano, R.P.; McCagg, A.; White, J.A.; Theroux, P.; Darius, H.; Lewis, B.S.; Ophuis, T.O.; Jukema, J.W.; *et al.* Ezetimibe Added to Statin Therapy after Acute Coronary Syndromes. *N. Engl. J. Med.* **2015**,*372*, 2387–2397.

32. Hung, H.-C.; Joshipura, K.J.; Jiang, R.; Hu, F.B.; Hunter, D.; Smith-Warner, S.A.; Colditz, G.A.; Rosner, B.; Spiegelman, D.; Willett, W.C. Fruit and vegetable intake and risk of major chronic disease. *J. Natl. Cancer Inst.* **2004**, *96*, 1577–1584.

33. Joshipura, K.J.; Hu, F.B.; Manson, J.E.; Stampfer, M.J.; Rimm, E.B.; Speizer, F.E.; Colditz, G.; Ascherio, A.; Rosner, B.; Spiegelman, D.; *et al.* The effect of fruit and vegetable intake on risk for coronary heart disease. *Ann. Intern. Med.* **2001**,*134*, 1106–1114.

34. Liu, S.; Manson, J.E.; Lee, I.-M.; Cole, S.R.; Hennekens, C.H.; Willett, W.C.; Buring, J.E. Fruit and vegetable intake and risk of cardiovascular disease: The Women's Health Study. *Am. J. Clin. Nutr.* **2000**, *72*, 922–928.

35. Liu, S.; Lee, I.-M.; Ajani, U. Intake of vegetables rich in carotenoids and risk of coronary heart disease in men: The Physicians' Health Study. *Int. J. Epidemiol.* **2001**, *30*, 130–135.

36. Bazzano, L.A.; He, J.; Ogden, L.G.; Loria, C.M.; Vupputuri, S.; Myers, L.; Whelton, P.K. Fruit and vegetable intake and risk of cardiovascular disease in US adults: The first National Health and Nutrition Examination Survey Epidemiologic Follow-up Study. *Am. J. Clin. Nutr.* **2002**, *76*, 93–99.

37. Jacobs, D.R.; Pereira, M.A.; Meyer, K.A.; Kushi, L.H. Fiber from whole grains, but not refined grains, is inversely associated with all-cause

mortality in older women: The Iowa Women's Health Study. *J. Am. Coll. Nutr.* **2000**, *19*, 326S–330S.

38. Eckel, R.H.; Jakicic, J.M.; Ard, J.D.; de Jesus, J.M.; Houston Miller, N.; Hubbard, V.S.; Lee, I.-M.; Lichenstein, A.H.; Loria, C.M.; Millen, B.E. A report of the American College of Cardiology/American Heart Association task force on practice guidelines. *Circulation* **2014**, *129*, S76–S99.

39. Yusuf, S.; Hawken, S.; Ounpuu, S.; Dans, T.; Avezum, A.; Lanas, F.; McQueen, M.; Budaj, A.; Pais, P.; Varigos, J.; *et al.* Effect of potentially modifiable risk factors associated with myocardial infarction in 52 countries (the INTERHEART study): Case-control study. *Lancet* **2004**, *364*, 937–952.

40. Mouritsen, O.G.; Zuckermann, M.J. What's so special about cholesterol? *Lipids* **2004**, *39*, 1101–1113.

41. Halling, K.K.; Slotte, J.P. Membrane properties of plant sterols in phospholipid bilayers as determined by differential scanning calorimetry, resonance energy transfer and detergent-induced solubilization. *Biochim. Biophys. Acta.* **2004**,*1664*, 161–171.

42. Devaraj, S.; Jialal, I. The role of dietary supplementation with plant sterols and stanols in the prevention of cardiovascular disease. *Nutr. Rev.* **2006**, *64*, 348–354.

43. Kris-Etherton, P.; Hecker, K.D.; Bonanome, A.; Coval, S.M.; Binkoski, A.E.; Hilpert, F.; Griel, A.E. Bioactive compounds in foods: Their role in the prevention of cardiovascular disease and cancer. *Am. J. Med.* **2002**, *113*, 71S–88S.

44. Katan, M.B.; Grundy, S.M.; Jones, P.; Law, M.; Miettinen, T.; Paoletti, R. Efficacy and safety of plant stanols and sterols in the management of blood cholesterol levels. *Mayo Clin. Proc.* **2003**, *78*, 965–978.

45. Nissinen, M.; Gylling, H.; Vuoristo, M.; Miettinen, T.A. Micellar distribution of cholesterol and phytosterols after duodenal plant stanol ester infusion. *Am. J. Physiol. Gastrointest. Liver Physiol.* **2002**, *282*, G1009–G1015.

46. Plat, J.; Mensink, R.P. Effects of plant stanol esters on LDL receptor protein expression and on LDL receptor and HMG-CoA reductase mRNA

expression in mononuclear blood cells of healthy men and women. *FASEB J.* **2002**, *16*, 258–260.

47. Jones, P.J.; Raeini-Sarjaz, M.; Ntanios, F.Y.; Vanstone, C.A.; Feng, J.Y.; Parsons, W.E. Modulation of plasma lipid levels and cholesterol kinetics by phytosterol *versus* phytostanol esters. *J. Lipid Res.* **2000**, *41*, 697–705.

48. The Food and Drug Administration (FDA). 65 FR 54686 - Food Labeling: Health Claims; Plant Sterols/Stanol Esters and Coronary Heart Disease. *Fed. Register* **2000**, *65*, 54686–54739.

49. The Food and Drug Administration (FDA). 75 FR 76526 - Food Labeling; Health Claim; Phytosterols and Risk of Coronary Heart Disease: Proposed Rule. *Fed. Register* **2010**, *75*, 76526–76571.

50. Niinikosk, I.H.; Viikari, J.; Palmmu, T. Cholesterol-lowering effect and sensory properties of sitostanol ester margarine in normocholesterolemic adults. *Scand. J. Nutr.* **1997**, *41*, 9–12.

51. Gylling, H.; Siimes, M.A.; Miettinen, T.A. Sitostanol ester margarine in dietary treatment of children with familial hypercholesterolemia. *J. Lipid Res.* **1995**, *36*, 1807–1812.

52. Gylling, H.; Miettinen, T.A. Serum cholesterol and cholesterol and lipoprotein metabolism in hypercholesterolaemic NIDDM patients before and during sitostanol ester-margarine treatment. *Diabetologia* **1994**, *37*, 773–780.

53. Volpe, R.; Niittynen, L.; Korpela, R.; Sirtori, C.; Bucci, A.; Fraone, N.; Pazzucconi, F. Effects of yoghurt enriched with plant sterols on serum lipids in patients with moderate hypercholesterolaemia. *Br. J. Nutr.* **2001**, *86*, 233–239.

54. Hendriks, H.F.; Weststrate, J.A.; van Vliet, T.; Meijer, G.W. Spreads enriched with three different levels of vegetable oil sterols and the degree of cholesterol lowering in normocholesterolaemic and mildly hypercholesterolaemic subjects.*Eur. J. Clin. Nutr.* **1999**, *53*, 319–327.

55. Pelletier, X.; Belbraouet, S.; Mirabel, D.; Mordret, F.; Perrin, J.L.; Pages, X.; Debry, G. A diet moderately enriched in phytosterols lowers plasma cholesterol concentrations in normocholesterolemic humans. *Ann. Nutr. Metab.* **1995**, *39*, 291–295.

56. Hendriks, H.F.; Brink, E.J.; Meijer, G.W.; Princen, H.M.; Ntanios, F.Y. Safety of long-term consumption of plant sterol esters-enriched spread. *Eur. J. Clin. Nutr.* **2003**, *57*, 681–692.

57. Miettinen, T.A.; Puska, P.; Gylling, H.; Vanhanen, H.; Vartiainen, E. Reduction of serum cholesterol with sitostanol-ester margarine in a mildly hypercholesterolemic population. *N. Engl. J. Med.* **1995**, *333*, 1308–1312.

58. Grundy, S.M. Stanol, Esters as a Component of Maximal Dietary Therapy in the National Cholesterol Education Program Adult Treatment Panel III Report. *Am. J. Cardiol.* **2005**, *96*, 47–50.

59. Han, X.; Shen, T.; Lou, H. Dietary polyphenols and their biological significance. *Int. J. Mol. Sci.* **2007**, *8*, 950–988.

60. Rice-Evans, C.A.; Miller, N.J.; Paganga, G. Structure-antioxidant activity relationships of flavonoids and phenolic acids. *Free Radic. Biol. Med.* **1996**, *20*, 933–956.

61. Yochum, L.; Kushi, L.H.; Meyer, K.; Folsom, A.R. Dietary flavonoid intake and risk of cardiovascular disease in postmenopausal women. *Am. J. Epidemiol.* **1999**, *149*, 943–949.

62. Knekt, P.; Jarvinen, R.; Reunanen, A.; Maatela, J. Flavonoid intake and coronary mortality in Finland: A cohort study.*Br. Med. J.* **1996**, *312*, 478–481.

63. Hertog, M.G.L.; Kromhout, D.; Aravanis, C.; Blackburn, H.; Buzina, R.; Fidanza, F.; Giampaoli, S.; Jansen, A.; Menotti, A.; Nedeljkovic, S.; *et al.* Flavonoid intake and long-term risk of coronary heart disease and cancer in the seven countries study. *Arch. Intern. Med.* **1995**, *155*, 381–386.

64. Hertog, M.G.L.; Feskens, E.J.M.; Hollman, P.C.H.; Katan, M.B.; Kromhout, D. Dietary antioxidant flavonoids and risk of coronary heart disease: The Zutphen Elderly Study. *Lancet* **1993**, *342*, 1007–1011.

65. Kammerer, D.; Claus, A.; Carle, R.; Schieber, A. Polyphenol screening of pomace from red and white grape varieties (Vitis vinifera L.) by HPLC-DAD-MS/MS. *J. Agric. Food Chem.* **2004**, *52*, 4360–4367.

66. Burns, J.; Gardner, P.T.; O'Neil, J.; Crawford, S.; Morecroft, I.; McPhail, D.B.; Lister, C.; Matthews, D.; MacLean, M.R.; Lean, M.E.; *et al.* Relationship among antioxidant activity, vasodilation capacity, and phenolic content of red wines. *J. Agric. Food Chem.* **2000**, *48*, 220–230.

67. Kerry, N.L.; Abbey, M. Red wine and fractionated phenolic compounds prepared from red wine inhibit low density lipoprotein oxidation *in vitro*. *Atherosclerosis* **1997**, *135*, 93–102.

68. Frankel, E.N.; Kanner, J.; German, J.B.; Parks, E.; Kinsella, J.E. Inhibition of oxidation of human low-density lipoprotein by phenolic substances in red wine. *Lancet* **1993**, *341*, 454–457.

69. Duthie, G.G.; Pedersen, M.W.; Gardner, P.T.; Morrice, P.C.; Jenkinson, A.M.; McPhail, D.B.; Steele, G.M. The effect of whisky and wine consumption on total phenol content and antioxidant capacity of plasma from healthy volunteers.*Eur. J. Clin. Nutr.* **1998**, *52*, 733–736.

70. Wollin, S.D.; Jones, P.J.H. Alcohol, red wine and cardiovascular disease. *J. Nutr.* **2001**, *131*, 1401–1404.

71. Rotondo, S.; de Gaetano, G. Protection from cardiovascular disease by wine and its derived products. Epidemiological evidence and biological mechanisms. *World Rev. Nutr. Diet.* **2000**, *87*, 90–113.

72. Freedman, J.E.; Parker, C.; Li, L.; Perlman, J.A.; Frei, B.; Ivanov, V.; Deak, L.R.; Iafrati, M.D.; Folts, J.D. Select flavonoid and whole juice from purple grapes inhibit platelet function and enhance nitric oxide release. *Circulation* **2001**, *103*, 2792–2798.

73. Xiang, L.; Xiao, L.; Wang, Y.; Li, H.; Huang, Z.; He, X. Health benefits of wine: Don't expect resveratrol too much. *Food Chem.* **2014**, *156*, 258–263.

74. Vlachogianni, I.C.; Fragopoulou, E.; Kostakis, I.K.; Antonopoulou, S. *In vitro* assessment of antioxidant activity of tyrosol, resveratrol and their acetylated derivatives. *Food Chem.* **2015**, *177*, 165–173.

75. Leonard, S.S.; Xia, C.; Jiang, B.H.; Stinefelt, B.; Klandorf, H.; Harris, G.K.; Shi, X. Resveratrol scavenges reactive oxygen species and effects radical-induced cellular responses. *Biochem. Biophys. Res. Commun.* **2003**, *309*, 1017–1026.

76. Pace-Asciak, C.R.; Hahn, S.; Diamandis, E.P.; Soleas, G.; Goldberg, D.M. The red wine phenolics trans-resveratrol and quercetin block human platelet aggregation and eicosanoid synthesis: Implications for protection against coronary heart disease. *Clin. Chim. Acta* **1995**, *235*, 207–219.

77. Subbaramaiah, K.; Chung, W.J.; Michaluart, P.; Telang, N.; Tanabe, T.; Inoue, H.; Jang, M.; Pezzuto, J.M.; Dannenberg, A.J. Resveratrol inhibits cyclooxygenase-2 transcription and activity in phorbol ester-treated human mammary epithelial cells. *J. Biol. Chem.* **1998**, *273*, 21875–21882.

78. Carluccio, M.A.; Siculella, L.; Ancora, M.A.; Massaro, M.; Scoditti, E.; Storelli, C.; Visioli, F.; Distante, A.; de Caterina, R. Olive oil and red wine antioxidant polyphenols inhibit endothelial activation: Antiatherogenic properties of Mediterranean diet phytochemicals. *Arterioscler. Thromb. Vasc. Biol.* **2003**, *23*, 622–629.

79. Li, Y.T.; Shen, F.; Liu, B.H.; Cheng, G.F. Resveratrol inhibits matrix metalloproteinase-9 transcription in U937 cells.*Acta Pharmacol. Sin.* **2003**, *24*, 1167–1171.

80. De la Lastra, C.A.; Villegas, I. Resveratrol as an anti-inflammatory and anti-aging agent: Mechanisms and clinical implications. *Mol. Nutr. Food Res.* **2005**, *49*, 405–430.

81. Hung, L.M.; Chen, J.K.; Huang, S.S.; Lee, R.S.; Su, M.J. Cardioprotective effect of resveratrol, a natural antioxidant derived from grapes. *Cardiovasc. Res.* **2000**, *47*, 549–555.

82. Szkudelski, T.; Szkudelska, K. Resveratrol and diabetes: From animal to human studies. *Biochem. Biophys. Acta* **2015**,*1852*, 1145–1154.

83. Fujitaka, K.; Otani, H.; Jo, F.; Jo, H.; Nomura, E.; Iwasaki, M.; Nishikawa, M.; Iwasaka, T.; Das, D.K. Modified resveratrol Longevinex improves endothelial function in adults with metabolic syndrome receiving standard treatment. *Nutr. Res.* **2011**, *31*, 842–847.

84. Wong, R.H.; Howe, P.R.; Buckley, J.D.; Coates, A.M.; Kunz, I.; Berry, N.M. Acute resveratrol supplementation improves flow-mediated dilatation in overweight/obese individuals with mildly elevated blood pressure. *Nutr. Metab. Cardiovasc. Dis.* **2011**, *21*, 851–856.

85. Wong, R.H.; Berry, N.; Coates, A.M.; Buckley, J.D.; Bryan, J.; Kunz, I.; Howe, P.R. Chronic resveratrol consumption improves brachial flow-mediated dilatation in healthy obese adults. *J. Hypertens.* **2013**, *31*, 1819–1827.

86. Agarwal, B.; Campen, M.J.; Channell, M.M.; Wherry, S.J.; Varamini, B.; Davis, J.G.; Baur, J.A.; Smoliga, J.M. Resveratrol for primary prevention of atherosclerosis: Clinical trial evidence for improved gene expression in vascular endothelium. *Int. J. Cardiol.* **2013**, *166*, 246–248.

87. Tome-Carneiro, J.; Gonzalvez, M.; Larrosa, M.; Yáñez-Gascón, M.J.; García-Almagro, F.J.; Ruiz-Ros, J.A.; García-Conesa, M.T.; Tomás-Barberán, F.A.; Espín, J.C. One-year consumption of a grape nutraceutical containing resveratrol improves the inflammatory and fibrinolytic status of patients in primary prevention of cardiovascular disease. *Am. J. Cardiol.* **2012**, *110*, 356–363.

88. Tome-Carneiro, J.; Gonzalvez, M.; Larrosa, M.; García-Almagro, F.J.; Avilés-Plaza, F.; Parra, S.; Yáñez-Gascón, M.J.; Ruiz-Ros, J.A.; García-Conesa, M.T.; Tomás-Barberán, F.A.; *et al.* Consumption of a grape extract supplement containing resveratrol decreases oxidized LDL and ApoB in patients undergoing primary prevention of cardiovascular disease: A triple-blind, 6-month follow-up, placebo-controlled, randomized trial. *Mol. Nutr. Food Res.***2012**, *56*, 810–821.

89. Goh, K.P.; Lee, H.Y.; Lau, D.P.; Supaat, W.; Chan, Y.H.; Koh, A.F. Effects of resveratrol in patients with type 2 diabetes mellitus on skeletal muscle SIRT1 expression and energy expenditure. *Int. J. Sport Nutr. Exerc. Metab.* **2014**, *24*, 2–13.

90. Bhatt, J.K.; Thomas, S.; Nanjan, M.J. Resveratrol supplementation improves glycemic control in type 2 diabetes mellitus. *Nutr. Res.* **2012**, *32*, 537–541.

91. Brasnyo, P.; Molnar, G.A.; Mohas, M.; Markó, L.; Laczy, B.; Cseh, J.; Mikolás, E.; Szijártó, I.A.; Mérei, A.; Halmai, R.; *et al.* Resveratrol improves insulin sensitivity, reduces oxidative stress and activates the Akt pathway in type 2 diabetic patients. *Br. J. Nutr.* **2011**, *106*, 383–389.

92. Muir, A.D.; Westcott, N.D. *Flax, the Genus Linum*; Taylor & Francis: London, UK, 2003.

93. Penalvo, J.L.; Haajanen, K.M.; Botting, N.; Adlercreutz, H. Quantification of lignans in food using isotope dilution gas chromatography/mass spectrometry. *J. Agric. Food Chem.* **2005**, *53*, 9342–9347.

94. Smeds, A.I.; Eklund, P.C.; Sjoholm, R.E.; Willfor, S.M.; Nishibe, S.; Deyama, T.; Holmbom, B.R. Quantification of a broad spectrum of lignans in cereals, oilseeds, and nuts. *J. Agric. Food Chem.* **2007**, *55*, 1337–1346.

95. Kuhnle., G.G.; Dell'Aaquila, C.; Aspinall., S.M.; Runswick., S.A.; Milligan, A.A.; Bingham, S.A. Phytoestrogen content of foods of animal origin: Dairy products, eggs, meat, fish, and seafood. *J. Agric. Food Chem.* **2008**, *56*, 10099–10104.

96. Rowland, I.; Faughnan, M.; Hoey, L.; Wahala, K.; Williamson, G.; Cassidy, A. Bioavailability of phyto-oestrogens. *Br. J. Nutr.* **2003**, *89*, S45–S58.

97. Setchell, K.D.; Lawson, A.M.; Borriello, S.P.; Harkness, R.; Gordon, H.; Morgan, D.M.; Kirk, D.N.; Adlercreutz, H.; Anderson, L.C.; Axelson, M. Lignan formation in man-microbial involvement and possible roles in relation to cancer.*Lancet* **1981**, *2*, 4–7.

98. Mousavi, Y.; Adlercreutz, H. Enterolactone and estradiol inhibit each other's proliferative effect on MCF-7 breast cancer cells in culture. *J. Steroid Biochem. Mol. Biol.* **1992**, *41*, 615–619.

99. Basly, J.P.; Lavier, M.C. Dietary phytoestrogens: Potential selective estrogen enzyme modulators? *Planta Med.* **2005**, *71*, 287–294.

100. Niemeyer, H.B.; Metzler, M. Differences in the antioxidant activity of plant and mammalian lignans. *J. Food Eng.* **2003**,*56*, 255–256.

101. Pellegrini, N.; Valtuena, S.; Ardigo, D.; Brighenti, F.; Franzini, L.; del Rio, D.; Scazzina, F.; Piatti, P.M.; Zavaroni, I. Intake of the plant lignans matairesinol, secoisolariciresinol, pinoresinol, and lariciresinol in relation to vascular inflammation and endothelial dysfunction in middle age-elderly men and post-menopausal women living in Northern Italy. *Nutr. Metab. Cardiovasc. Dis.* **2010**, *20*, 64–71.

102. Englyst, H.; Wiggins, H.S.; Cummings, J.H. Determination of the non-starch polysaccharides in plant foods by gas-liquid

chromatography of constituent sugars as alditol acetates. *Analyst* **1982**, *107*, 307–318.

103. Englyst, H.N.; Kingman, S.; Cummings, J. Classification and measurement of nutritionally important starch fractions.*Eur. J. Clin. Nutr.* **1992**, *46*, S33–S50.

104. Goldring, J.M. Resistant starch: Safe intakes and legal status. *J. AOAC Int.* **2004**, *87*, 733–739.

105. Fuentez-Zaragoza, E.; Riquelme-Navarrete, M.J.; Sanchez-Zapta, E.; Perez-Alvarez, J.A. Resistant starch as a functional ingredient: A review. *Food Res. Int.* **2010**, *43*, 931–942.

106. Nichenametia, S.N.; Weidauer, L.A.; Wey, H.E.; Beare, T.M.; Specker, B.L.; Dey, M. Resistant starch type 4-enriched diet lowered blood cholesterols and improved body composition in a double blind controlled cross-over intervention.*Mol. Nutr. Food Res.* **2014**, *58*, 1365–1369.

107. Haub, M.D.; Hubach, K.L.; Al-Tamimi, E.K.; Ornelas, S.; Seib, P.A. Different types of resistant starch elicit different glucose responses. *J. Nutr. Metab.* **2010**.

108. Maki, K.C.; Phillips, A.K. Dietary substitutions for refined carbohydrate that show promise for reducing risk of type 2 diabetes in men and women. *J. Nutr.* **2015**, *145*, 159S–163S.

109. Birt, D.F.; Boylston, T.; Hendrich, S.; Jane, J.L.; Hollis, J.; Li, L.; McClelland, J.; Moore, S.; Phillips, G.J.; Rowling, M.; *et al.* Resistant starch: Promise for improving human health. *Adv. Nutr.* **2013**, *4*, 587–601.

110. Robertson, M.D.; Wright, J.W.; Loizon, E.; Debard, C.; Vidal, H.; Shojaee-Moradie, F.; Umpleby, A.M. Insulin-sensitizing effects on muscle and adipose tissue after dietary fiber intake in men and women with metabolic syndrome. *Am. J. Clin. Nutr.* **2012**, *82*, 559–567.

111. Johnston, K.; Thomas, E.; Bell, J.; Frost, G.; Robertson, M. Resistant starch improves insulin sensitivity in metabolic syndrome. *Diabet. Med.* **2010**, *27*, 391–397.

112. Robertson, M.D.; Bickerton, A.S.; Dennis, A.L.; Vidal, H.; Frayn, K.N. Insulin-sensitizing effects of dietary resistant starch and effects on skeletal muscle and adipose tissue metabolism. *Am. J. Clin. Nutr.* **2005**, *82*, 559–567.

113. Maki, K.C.; Pelkman, C.L.; Finocchiaro, E.T.; Kelley, K.M.; Lawless, A.L.; Schild, A.L.; Rains, T.M. Resistant starch from high-amylose maize increases insulin sensitivity in overweight and obese men. *J. Nutr.* **2012**, *142*, 717–723.

114. Bodinham, C.L.; Smith, L.; Thomas, E.L.; Bell, J.D.; Swann, J.R.; Costabile, A.; Russell-Jones, D.; Umpleby, A.M.; Robertson, M.D. Efficacy of increased resistant starch consumption in human type 2 diabetes. *Endocr. Connect.* **2014**, *3*, 75–84.

115. Brownawell, A.M.; Caers, W.; Gibson, G.R.; Kendall, C.W.; Lewis, K.D.; Ringel, Y.; Slavin, J.L. Prebiotics and the health benefits of fiber: Current regulatory status, future research, and goals. *J. Nutr.* **2012**, *142*, 962–974.

116. Lin, H.V.; Frassetto, A.; Kowalik, E.J., Jr.; Nawrocki, A.R.; Lu, M.M.; Kosinski, J.R.; Forrest, G. Butyrate and propionate protect against diet-induced obesity and regulate gut hormones via free fatty acid receptor 3-independent mechanisms. *PLoS ONE* **2012**, *7*, e35240.

117. Burcelin, R. The incretins: A link between nutrients and well-being. *Br. J. Nutr.* **2005**, *93*, S147–S156.

118. Batterham, R.L.; Bloom, R.L. The gut hormone peptide YY regulates appetite. *Ann. N. Y. Acad. Sci.* **2003**, *994*, 162–168.

119. Zhang, L.; Hua Ting, L.I.; Shen, L.; Fang, Q.C.; Qian, L.L.; Wei Ping, J.I.A. Effect of dietary resistant starch on the prevention and treatment of obesity-related diseases and its possible mechanisms. *Biomed. Environ. Sci.* **2015**, *28*, 291–297.

120. Zhou, J.; Martin, R.J.; Raggio, A.M.; Shen, L.; McCutcheon, K.; Keenan, M.J. The importance of GLP-1 and PYY in resistant starch's effect on body fat in mice. *Mol. Nutr. Food Res.* **2015**, *59*, 1000–1003.

121. Higgins, J.A.; Brown, M.A.; Storlien, L.H. Consumption of resistant starch decreases postprandial lipogenesis in white adipose tissue of the rat. *Nutr. J.* **2006**, *5*, 1–4.

122. Aziz, A.A.; Kenney, L.S.; Goulet, B.; Abdel-Aal, E. Dietary starch type affects body weight and glycemic control in freely fed but not energy-restricted obese rats. *J. Nutr.* **2009**, *139*, 1881–1889.

123. Higgins, J.A. Resistant starch and energy balance: Impact on weight loss and maintenance. *Crit. Rev. Food Sci. Nutr.***2014**, *54*, 1158–1166.

124. Robertson, M.D. Dietary resistant starch and glucose metabolism. *Curr. Opin. Clin. Nutr. Metab. Care* **2012**, *15*, 362–367.

125. Zhang, C.; Zhang, M.; Wang, S.; Han, R.; Cao, Y.; Hua, W.; Mao, Y.; Zhang, X.; Pang, X.; Wei, C.; *et al.* Interactions between gut microbiota, host genetics and diet relevant to development of metabolic syndromes in mice. *ISME J.***2010**, *4*, 232–241.

126. García-Rodríguez, C.E.; Mesa, M.D.; Olza, J.; Buccianti, G.; Pérez, M.; Moreno-Torres, R.; de la Cruz, A.P.; Gil, A. Postprandial glucose, insulin and gastrointestinal hormones in healthy and diabetic subjects fed a fructose-free and resistant starch type IV-enriched enteral formula. *Euro. J. Nutr.* **2013**, *52*, 1569–1578.

127. Hui, H.; Farilla, L.; Merkel, P.; Perfetti, R. The short half-life of glucagon-like peptide-1 in plasma does not reflect its long-lasting beneficial effects. *Eur. J. Endocrinol.* **2002**, *146*, 863–869.

128. Behall, K.M.; Howe, J.C. Resistant starch as energy. *J. Am. Coll. Nutr.* **1996**, *15*, 248–254.

129. Le Leu, R.K.; Winter, J.M.; Christophersen, C.T.; Young, G.P.; Humphreys, K.J.; Hu, Y.; Gratz, S.W.; Miller, R.B.; Topping, D.L.; Bird, A.R.; *et al.* Butyrylated starch intake can prevent red meat-induced O6-methyl-2-deoxyguanosine adducts in human rectal tissue: A randomized clinical trial. *Br. J. Nutr.* **2015**, *114*, 220–230.

130. Vaziri, N.D.; Liu, S.M.; Lau, W.L.; Khazaeli, M.; Nazertehrani, S.; Farzaneh, S.H.; Kieffer, D.A.; Adams, S.H.; Martin, R.J. High amylose resistant starch diet ameliorates oxidative stress, inflammation, and progression of chronic kidney disease. *PLoS ONE* **2014**, *9*, e114881.

131. Curtius, T.; Goebel, F. Uber Glycollather. *J. Prakt. Chem.* **1988**, *37*, 150–181.

132. Abderhalden, E.; Haas, R. Further studies on the structure of proteins: Studies on the physical and chemical properties of 2,5-di-ketopiperazines. *Z. Physiol. Chem.* **1926**, *151*, 114–l19.

133. Prasad, C. Bioactive cyclic dipeptides. *Peptides* **1995**, *16*, 151–164.

134. Prasad, C. Neurobiology of cyclo (His-Pro). *Ann. N. Y. Acad. Sci.* **1989**, *553*, 232–251.

135. Prasad, C.; Kumar, S.; Adkinson, W.; McGregor, J.U. Hormones in foods: Abundance of authentic cyclo (His-Pro)-like immunoreactivity in milk and yogurt. *Nutr. Res.* **1995**, *15*, 1623–1635.

136. Hilton, C.W.; Prasad, C.; Vo, P.; Mouton, C. Food contains the bioactive peptide, cyclo (His-Pro). *J. Clin. Endocrinol. Metab.* **1992**, *75*, 375–378.

137. Hilton, C.W.; Prasad, C.; Svec, F.; Vo, P.; Reddy, S. Cyclo (His-Pro) in nutritional supplements. *Lancet* **1990**, *336*, 1455.

138. Mori, M.; Prasad, C.; Wilber, J.F. Specific radioimmunoassay of cyclo (His-Pro), a biologically active metabolite of thyrotropin-releasing hormone. *Endocrinology* **1981**, *108*, 1995–1997.

139. Bellezza, I.; Peirce, M.J.; Minelli, A. Cyclic dipeptides: From bugs to brain. *Trends Mol. Med.* **2014**, *20*, 551–558.

140. Minelli, A.; Bellezza, I.; Grottelli, S.; Galli, F. Focus on cyclo (His-Pro): History and perspectives as antioxidant peptide. *Amino Acids* **2008**, *35*, 283–289.

141. Prasad, C. Cyclo (His-Pro): Its distribution, origin and function in the human. *Neurosci. Biobehav. Rev.* **1988**, *12*, 19–22.

142. Morley, J.E.; Levine, A.S.; Prasad, C. Histidyl-proline diketopiperazine decreases food intake in rats. *Brain Res.* **1981**,*210*, 475–478.

143. Kow, L.M.; Pfaff, D.W. The effects of the TRH metabolite cyclo (His-Pro) and its analogs on feeding. *Pharmacol. Biochem. Behav.* **1991**, *38*, 359–364.

144. Kow, L.M.; Pfaff, D.W. Cyclo (His-Pro) potentiates the reduction of food intake induced by amphetamine, fenfluramine, or serotonin. *Pharmacol. Biochem. Behav.* **1991**, *38*, 365–369.

145. Wilber, J.F.; Mori, M.; Pegues, J.; Prasad, C. Endogenous histidyl-proline diketopiperazine [cyclo (His-Pro)]: A potential satiety neuropeptide in normal and genetically obese rodents. *Trans. Assoc. Am. Physicians* **1983**, *96*, 131–136.

146. Bowden, C.R.; Karkanias, C.D.; Bean, A.J. Re-evaluation of histidyl-proline diketopiperazine [cyclo (His-Pro)] effects on food intake in the rat. *Pharmacol. Biochem. Behav.* **1988**, *29*, 357–363.

147. Prasad, C.; Mizuma, H.; Brock, J.W.; Porter, J.R.; Svec, F.; Hilton, C. A paradoxical elevation of brain cyclo (His-Pro) levels in hyperphagic obese Zucker rats. *Brain Res.* **1995**, *699*, 149–153.

148. Mori, M.; Pegues, J.; Prasad, C.; Wilber, J.F. Fasting and feeding-associated changes in cyclo (His-Pro)-like immunoreactivity in the rat brain. *Brain Res.* **1983**, *268*, 181–184.

149. Steiner, H.; Wilber, J.F.; Prasad, C.; Rogers, D.; Rosenkranz, R.T. Histidyl proline diketopiperazine (Cyclo [His-Pro]) in eating disorders. *Neuropeptides* **1989**, *14*, 185–189.

150. Song, M.K.; Hwang, I.K.; Rosenthal, M.J.; Harris, D.M.; Yamaguchi, D.T.; Yip, I.; Go, V.L. Anti-Hyperglycemic Activity of Zinc Plus Cyclo (His-Pro) in Genetically Diabetic Goto-Kakizaki and Aged Rats. *Exp. Biol. Med.* **2003**, *228*, 1338–1345.

151. Song, M.K.; Rosenthal, M.J.; Song, A.M.; Uyemura, K.; Yang, H.; Ament, M.E.; Yamaguchi, D.T.; Cornford, E.M. Body weight reduction in rats by oral treatment with zinc plus cyclo-(His-Pro). *Br. J. Pharmacol.* **2009**, *158*, 442–450.

152. Mori, M.; Pegues, J.; Prasad, C.; Wilber, J.; Peterson, J.; Githens, S. Histidyl-proline diketopiperazine cyclo (His-Pro): Identification and characterization in rat pancreatic islets. *Biochem. Biophys. Res. Commun.* **1983**, *115*, 281–286.

153. Leduque, P.; Jackson, I.M.; Kervran, A.; Aratan-Spire, S.; Czernichow, P.; Dubois, P.M. Histidyl-Proline Diketopiperazine (His-Pro

DKP) Immunoreactivity is Present in the Glucagon-Containing Cells of the Human Fetal Pancreas. *J. Clin. Investig.* **1987**, *79*, 875–880.

154. Hilton, C.W.; Prasad, C.; Wilber, J.F. Acute alterations of cyclo (His-Pro) levels after oral ingestion of glucose.*Neuropeptides* **1990**, *15*, 55–59.

155. Hilton, C.W.; Reddy, S.; Prasad, C.; Wilber, J.F. Change in circulating cyclo (His-Pro) concentrations in rats after ingestion of oral glucose compared to intravenous glucose and controls. *Endocr. Res.* **1990**, *16*, 139–150.

156. Song, M.K.; Rosenthal, M.J.; Song, A.M.; Yang, H.; Ao, Y.; Yamaguchi, D.T. Raw vegetable food containing high cyclo (his-pro) improved insulin sensitivity and body weight control. *Metabolism* **2005**, *54*, 1480–1489.

157. Jung, E.Y.; Lee, H.S.; Choi, J.W.; Ra, K.S.; Kim, M.R.; Suh, H.J. Glucose tolerance and antioxidant activity of spent brewer's yeast hydrolysate with a high content of Cyclo-His-Pro (CHP). *J. Food Sci.* **2011**, *76*, C272–C278.

158. Hilton, C.W.; Mizuma, H.; Svec, F.; Prasad, C. Relationship between plasma cyclo (His-Pro), a neuropeptide common to processed protein-rich food, and C-peptide/insulin molar ratio in obese women. *Nutr. Neurosci.* **2001**, *4*, 469–474.

159. Samols, E.; Marri, G.; Marks, V. Promotion of insulin secretion by glucagon. *Lancet* **1965**, *2*, 415–416.

160. Huypens, P.; Ling, Z.; Pipeleers, D.; Schuit, F. Glucagon receptors on human islet cells contribute to glucose competence of insulin release. *Diabetologia* **2000**, *43*, 1012–1019.

161. Kawai, K.; Yokota, C.; Ohashi, S.; Watanabe, Y.; Yamashita, K. Evidence that glucagon stimulates insulin secretion through its own receptor in rats. *Diabetologia* **1995**, *38*, 274–276.

162. Pivovarova, O.; Bernigau, W.; Bobbert, T.; Isken, F.; Möhlig, M.; Spranger, J.; Weickert, M.O.; Osterhoff, M.; Pfeiffer, A.F.; Rudovich, N. Hepatic insulin clearance is closely related to metabolic syndrome components. *Diabetes Care* **2013**,*36*, 3779–3785.

163. Meshkani, R.; Adeli, K. Hepatic insulin resistance, metabolic syndrome and cardiovascular disease. *Clin. Biochem.***2009**, *42*, 1331–1346.

164. Biddinger, S.B.; Hernandez-Ono, A.; Rask-Madsen, C.; Haas, J.T.; Alemán, J.O.; Suzuki, R.; Scapa, E.F.; Agarwal, C.; Carey, M.C.; Stephanopoulos, G.; *et al*. Hepatic insulin resistance is sufficient to produce dyslipidemia and susceptibility to atherosclerosis. *Cell Metab.* **2008**, *7*, 125–134.

165. Basu, A.; Betts, N.M.; Nguyen, A.; Newman, E.D.; Fu, D.; Lyons, T.J. Freeze-dried strawberries lower serum cholesterol and lipid peroxidation in adults with abdominal adiposity and elevated serum lipids. *J. Nutr.* **2014**, *144*, 830–837.

166. Leong, D.J.; Choudhury, M.; Hirsh, D.M.; Hardin, J.A.; Cobelli, N.J.; Sun, H.B. Nutraceuticals: Potential for Chondroprotection and Molecular Targeting of Osteoarthritis. *Int. J. Mol. Sci.* **2013**, *14*, 23063–23085.

167. Huntley, A.L. The health benefits of berry flavonoids for menopausal women: Cardiovascular disease, cancer and cognition. *Maturitas* **2009**, *63*, 297–301.

168. Szajdek, A.; Borowska, E.J. Bioactive compounds and health-promoting properties of berry fruits: A review. *Plant Foods Hum. Nutr.* **2008**, *63*, 147–156.

169. Barnes, S. Nutritional genomics, polyphenols, diets, and their impact on dietetics. *J. Am. Diet. Assoc.* **2008**, *108*, 1888–1895.

170. Shen, C.L.; Smith, B.J.; Lo, D.-F.; Chyu, M.-C.; Dunn, D.M.; Chen, C.-H.; Kwun, I.-S. Dietary polyphenols and mechanisms of osteoarthritis. *J. Nutr. Biochem.* **2012**, *23*, 1367–1377.

171. Alvarez-Suarez, J.M.; Giampieri, F.; Tulipani, S.; Casoli, T.; di Stefano, G.; Gonzalez-Paramas, A.M.; Santos-Buelga, C.; Busco, F.; Quiles, J.L.; Cordero, M.D.; *et al*. One-month strawberry-rich anthocyanin supplementation ameliorates cardiovascular risk, oxidative stress markers and platelet activation in humans. *J. Nutr. Biochem.* **2014**, *25*, 289–294.

172. Rodriguez-Mateos, A.; Ishisaka, A.; Mawatari, K.; Vidal-Diez, A.; Spencer, J.P.E.; Terao, J. Blueberry intervention improves vascular reactivity and lowers blood pressure in high-fat-, high-cholesterol-fed rats. *Br. J. Nutr.* **2013**, *109*, 1746–1754.

173. Zunino, S.J.; Parelman, M.A.; Freytag, T.L.; Stephensen, C.B.; Kelley, D.S.; Mackey, B.E.; Woodhouse, L.R.; Bonnel, E.L. Effects of dietary strawberry powder on blood lipids and inflammatory markers in obese human subjects. *Br. J. Nutr.* **2012**, *108*, 900–909.

174. Basu, A.; Du, M.; Leyva, M.J.; Sanchez, K.; Betts, N.M.; Wu, M.; Aston, C.E.; Lyons, T.J. Blueberries decrease cardiovascular risk factors in obese men and women with metabolic syndrome. *J. Nutr.* **2010**, *140*, 1582–1587.

175. Erlund, I.; Koli, R.; Alfhan, G.; Marniemi, J.; Puukka, P.; Mustonen, P.; Mattila, P.; Jula, A. Favorable effects of berry consumption on platelet function, blood pressure and HDL cholesterol. *Am. J. Clin. Nutr.* **2008**, *87*, 323–331.

176. Ruel, G.; Pomerleau, S.; Couture, P.; Lemieux, S.; Lamarche, B.; Couillard, C. Favourable impact of low-calorie cranberry juice consumption on plasma HDL-cholesterol concentrations in men. *Br. J. Nutr.* **2006**, *96*, 354–364.

177. Ruel, G.; Pomerleau, S.; Couture, P.; Lamarche, B.; Couillard, C. Changes in plasma antioxidant capacity and oxidized low-density lipoprotein levels in men after short-term cranberry juice consumption. *Metabolism* **2005**, *54*, 856–861.

178. Rodriguez-Mateos, A.; Rendeiro, C.; Bergillos-Mecca, T.; Tabatabaee, S.; George, T.W.; Heiss, C.; Spencer, J.P. Intake and time dependence of blueberry flavonoid-induced improvements in vascular function: A randomized, controlled, double-blind, crossover intervention study with mechanistic insights into biological activity. *Am. J. Clin. Nutr.* **2013**,*98*, 1179–1191.

179. Heinonen, I.M.; Meyer, A.S.; Frankel, E.N. Antioxidant activity of berry polyphenolics on human low-density lipoprotein and liposome oxidation. *J. Agric. Food Chem.* **1998**, *46*, 4107–4112.

180. Meyer, A.S.; Donovan, J.L.; Pearson, D.A.; Waterhouse, A.L.; Frankel, E.N. Fruit hydroxycinnamic acids inhibit human low-density

lipoprotein oxidation *in vitro*. *J. Agric. Food Chem.* **1998**, *46*, 1783–1787.

181. Coban, J.; Evran, B.; Ozkan, F.; Cevik, A.; Dogru-Abbasoglu, S.; Uysal, M. Effect of blueberry feeding on lipids and oxidative stress in the serum, liver and aorta of Guinea pigs fed on a high-cholesterol diet. *Biosci. Biotechnol. Biochem.***2013**, *77*, 389–391.

182. Del Bo, C.; Riso, P.; Campolo, J.; Moller, P.; Loft, S.; Klimis-Zacas, D.; Brambilla, A.; Rizzolo, A.; Porrini, M. A single portion of blueberry (*Vaccinium corymbosum* L.) improves protection against DNA damage but not vascular function in healthy male volunteers. *Nutr. Res.* **2013**, *33*, 220–227.

183. Cassidy, A.; O-Reilly, E.; Kay, C.; Sampson, L.; Franz, M.; Forman, J.P.; Curhan, G.; Rimm, E.B. Habitual intake of flavonoid subclasses and incident hypertension in adults. *Am. J. Clin. Nutr.* **2011**, *93*, 338–347.

184. Johnson, S.A.; Figueroa, A.; Navaei, N.; Wong, A.; Kalfon, R.; Ormsbee, L.T.; Feresin, R.G.; Elam, M.L.; Hooshmand, S.; Payton, M.E.; *et al.* Daily blueberry consumption improves blood pressure and arterial stiffness in postmenopausal women with pre-and stage-1 hypertension: A randomized, double-blind, placebo-controlled clinical trial. *J. Am. Diet. Assoc.* **2015**, *115*, 369–377.

185. McAnulty, L.S.; Collier, S.R.; Landram, M.J.; Whittaker, D.S.; Isaacs, S.E.; Klemka, J.M.; Cheek, S.L.; Arms, J.C.; McAnulty, S.R. Six weeks daily ingestion of whole blueberry powder increases natural killer cell counts and reduces arterial stiffness in sedentary males and females. *Nutr. Res.* **2014**, *34*, 577–584.

186. McAnulty, S.R.; McAnulty, L.S.; Morrow, J.D.; Khardouni, D.; Shooter, L.; Monk, J.; Gross, S.; Brown, V. Effect of daily fruit ingestion on angiotensin converting enzyme activity, blood pressure, and oxidative stress in chronic smokers.*Free Radic. Res.* **2005**, *39*, 1241–1248.

187. Stull, A.J.; Cash, K.C.; Champagne, C.M.; Gupta, A.K.; Boston, R.; Beyl, R.A.; Johnson, W.D.; Cefalu, W.T. Blueberries improve endothelial function, but not blood pressure, in adults with metabolic syndrome: A randomized, double-blind, placebo-controlled clinical trial. *Nutrients* **2015**, *7*, 4107–4123.

188. Amani, R.; Moazen, S.; Shahbazian, H.; Ahmadi, K.; Jalali, M.T. Flavonoid-rich beverage effects on lipid profile and blood pressure in diabetic patients. *World J. Diabetes* **2014**, *5*, 962–968.

189. Martineau, L.C.; Couture, A.; Spoor, D.; Benhaddou-Andaloussi, A.; Harris, C.; Meddah, B.; Leduc, C.; Burt, A.; Vuong, T.; Mai Le, P.; *et al.* Anti-diabetic properties of the Canadian lowbush blueberry Vaccinium angustifolium Ait.*Phytomedicine* **2006**, *13*, 612–623.

190. Jayaprakasam, B.; Vareed, S.K.; Olson, L.K.; Nair, M.G. Insulin secretion by bioactive anthocyanins and anthocyanidins present in fruits. *J. Agric. Food Chem.* **2005**, *53*, 28–31.

191. Stull, A.J.; Cash, K.C.; Johnson, W.D.; Champagne, C.M.; Cefalu, W.T. Bioactives in blueberries improve insulin sensitivity in obese, insulin-resistant men and women. *J. Nutr.* **2010**, *140*, 1764–1768.

192. DeFuria, J.; Bennett, G.; Strissel, K.J.; Perfield, J.W., II; Milbury, P.E.; Greenberg, A.S.; Obin, M.S. Dietary blueberry attenuates whole-body insulin resistance in high fat-fed mice by reducing adipocyte death and its inflammatory sequelae. *J. Nutr.* **2009**, *139*, 1510–1516.

193. Shahidi, F.; Naczk, M. Food phenolics: An overview. In *Food Phenolics: Sources, Chemistry, Effects, Applications*; Shahidi, F., Naczk, M., Eds.; Technomic Publishing Company Inc.: Lancaster, PA, USA, 1995; pp. 1–5.

194. Hennekens, C.H.; Buring, J.E.; Manson, J.E.; Stampfer, M.; Rosner, B.; Cook, N.R.; Belanger, C.; LaMotte, F.; Gaziano, J.M.; Ridker, P.M.; *et al.* Lack of effect of long-term supplementation with β-carotene on the incidence of malignant neoplasms and cardiovascular disease. *N. Engl. J. Med.* **1996**, *334*, 1145–1149.

Chapter 7

Effect of Chitosan on Disease Control and Growth of Scots Pine (*Pinus sylvestris*L.) in a Forest Nursery

Marta Aleksandrowicz-Trzcińska [1,*], Anna Bogusiewicz [1], Michał Szkop [2] and Stanisław Drozdowski [3]

[1]Department of Forest Protection and Ecology, Faculty of Forestry, Warsaw University of Life Sciences, Nowoursynowska 159, Warsaw 02-776, Poland

[2]Department of Biochemistry, Faculty of Agriculture and Biology, Warsaw University of Life Sciences, Nowoursynowska 159, Warsaw 02-776, Poland

[3]Department of Silviculture, Faculty of Forestry, Warsaw University of Life Sciences, Nowoursynowska 159, Warsaw 02-776, Poland

ABSTRACT

Chitosan has become a promising biological agent for disease control and plant growth promotion. The objective of this study was to assess the effects of chitosan, applied as an active ingredient of Beta-chikol (Poli-Farm, Łowicz, Poland), to control damping-off and *Lophodermium* needle cast on Scots pine seedlings growing in field conditions. Beta-chikol was used for seed treatment and as a foliar spray at recommended rates and concentrations. For each experimental variant (chitosan, fungicides, unprotected), inventories of seedlings were performed, after germination and again after six weeks. In the aboveground parts of seedlings, the concentration of endogenous salicylic acid was determined by HPLC. At the end of the growing season, seedling growth parameters were determined. Beta-chikol used as foliar spray limited infection by the damping-off fungi but was ineffective when used as a seed treatment.*Lophodermium* needle cast was not observed during the study period. After the application of Beta-chikol, the concentration of salicylic acid did not increase. The application of Beta-chikol enhanced all growth parameters under investigation. Our results indicate the possibility of using chitosan in the form of Beta-chikol to stimulate plant growth and protect pine seedlings against parasitic damping-off in forest nurseries.

KEYWORDS

Beta-chikol; chitosan; *Pinus sylvestris*; salicylic acid; growth stimulation; disease control

1. INTRODUCTION

Polymers have been used worldwide for the last few decades. They have a number of applications, ranging from daily needs to biomedical and defence fields [1,2,3,4,5]. Among various polymers, biorenewable polymers such as cellulose, hemicellulose, lignin, and chitosan have attracted the attention of researchers and manufactures as low cost materials for several industrial applications [2,3,4,5,6]. Among these biorenewable polymers, polysaccharides are the most abundant [3,4,5,6]. Their unique properties include flexibility, biodegradability, environmental friendliness, easy processing, recyclability, impressive physico-machanical properties, and easy availability [5]. Chitosan is one of the most important natural polymers [6].

The intensive production of seedlings creates favorable conditions for the development of fungal pathogens, so it is necessary to use plant protection products in forest nurseries. Whenever possible, the use of chemicals should be minimized, while alternative substances and biological methods should be preferred. One such substance is chitosan [7,8].

Chitosan is an organic polysaccharide also known as deacetylchitin, as it is a deacetylated derivative of chitin. On a large scale, it is produced from the skeletons of shrimps and crabs. Its advantages include non-toxicity, biodegradability, and high bioactivity, as well as sorption and chelating properties [9].

Chitosan is an exogenous elicitor of plant defense responses. It induces local and systemic acquired resistance, reflected in the synthesis of salicylic acid, phytoalexins, and pathogenesis-related proteins (chitinase and β-1, 3-glucanase), the lignification of cell walls, and callose synthesis [8,10,11,12]. In seeds treated with chitosan, increasing concentrations of phenolic compounds—derivatives of hydroxycinnamic acid (*p*-cumaric, caffeic and feluric) and benzoic acid (benzoic, pirotocatechulic and gallic)—increase seed resistance to infection due to lignin synthesis and accumulation [13,14]. Chitosan affects plant defences in two ways, as it not only activates the genes responsible for the initiation of resistance mechanisms in plants, but it also has antiviral [15], antibacterial [9,16], and antifungal properties [17,18].

Moreover, chitosan stimulates plant growth and development. Seed treatment with chitosan promotes germination, increases seedling emergence, and accelerates further plant growth [19,20]. Many studies have shown that plants treated with chitosan are taller and have larger and more numerous leaves. Chitosan application also shortens the vegetative phase of plants and accelerates flowering; the treated plants produce more numerous and larger flowers. In addition, crop yields are increased [21].

Salicylic acid is a phenolic compound that is a derivative of benzoic acid commonly found in plants in small concentrations, below 1 mg/kg fresh weight [22]. However, in infected plants its concentration can increase 20-fold, activating the genes responsible for the synthesis of defense-related proteins [23]. Both endogenous and exogenous salicylic acid induces local resistance and is a signal molecule for the development of systemic acquired resistance [24]. Salicylic acid increases plant tolerance to abiotic stressors: low temperature, soil salinity, and herbicide treatment [25,26,27]. Moreover, salicylic acid is an endogenous regulator of plant growth and development [28]. There are also studies that show a positive effect of exogenous salicylic acid on aboveground and root dry mass, the regeneration of flower buds, flowering, and tuber formation [26,29].

Research on chitosan as a plant protection product and growth stimulant has so far concerned various herbaceous crop plants, while only few studies have focused on woody plants, including forest trees [8,21,30]. This study aimed to assess the effectiveness of chitosan applied in the form of Beta-chikol as a seed treatment agent and foliar spray to protect Scots pine (*Pinus sylvestris* L.) against parasitic damping-off and *Lophodermium* needle cast in a forest nursery (in the ground). In the experiment, we also assessed the effect of this bioproduct on plant growth. Our null hypotheses were that (1) Beta-chikol limits seedling infection from both diseases; and (2) it has a favorable effect on pine growth. We also assumed that the major compound that determines the effectiveness of chitosan as a stimulant of pine resistance and growth is endogenous salicylic acid, whose concentration would increase as a result of the Beta-chikol treatment.

2. MATERIALS AND METHODS

2.1. Study Site

Field research was conducted in the forest nursery of the Spała Forest District (51.54121 N; 20.14754 E), located approximately 120 km SW of Warsaw in Poland. The nursery started to produce tree seedlings in 1958. The annual mean temperature in the study area is 7.2 °C. The warmest month is July (17.3 °C), and the coldest month is January (−3.2 °C). The growing season (mean daily air temperature above +5 °C) begins in early April and lasts on average 210 days. The long-term mean annual precipitation is 596 mm, of which approximately 420 mm fall during the growing season [31].

2.2. Filed Experiment, Application of Chitosan and Fungicide

The experiment compared 3 variants: pine seedlings protected with either chitosan or fungicides against parasitic damping-off and *Lophodermium* needle cast and unprotected seedlings. Seeds were sown on 8 May, 2013, because of excessive precipitation earlier in the year. Chitosan was applied in the form of Beta-chikol (Poli-Farm, Łowicz, Poland) as an organic plant growth stimulant. This chitosan formulation is in liquid form for dilution with water. The concentration of chitosan in the product is 2%. We used it for seed priming (soaking before sowing) and as a foliar spray at a concentration of 2%. The seeds were soaked in the solution for 5 hours before sowing. For spraying, 1000 L of the solution were used per hectare.

The fungicide treatment involved a dry seed treatment with Sadoplon 75 WP and spraying with Previcur Energy 840 SL, Gwarant 500 SC, Amistar 250 SC, and Falcon 460 EC. The fungicides were applied according to the standard procedures in Poland, defined by the Forest Research Institute [32].

Chitosan was applied 3 times after seed germination: on 7 June, 12 July, and 26 August. Fungicides were applied on the same dates and an additional 8 times (11 in total: for protection against damping-off and needle cast). These included 6 treatments every 5–7 days against parasitic damping-off (alternating Previcur Energy and Gwarant) and 5 treatments every 3–4 weeks against needle-cast (alternating Amistar and Falcon).

2.3. Seedling Inventory and Growth Measurement

Seedling emergence was estimated twice, on the days of the 1st and 2nd chitosan treatments. Seedlings were counted in 1-metre segments of individual rows of the seedbed, spaced 5 m apart. For each variant, 32 such 1-metre segments were taken into account. In late October, 60 seedlings of each variant were collected to measure biometric characters. For each seedling, we measured the shoot length, root collar diameter, dry mass of aboveground parts and roots, and we counted lateral shoots. Using the WinRHIZO software (Regent Instruments Inc., Québec, Canada), the root length and number of root tips were estimated.

2.4. Salicylic Acid Content

To measure the salicylic acid concentration, seedlings were sampled twice: 5 and 12 days after the 2nd chitosan treatment. A single sample consisted of 10 seedlings. In total, 18 samples were taken, 3 for each variant on the 2 dates.

The salicylic acid content was determined using reversed-phase–high-performance liquid chromatography (RP-HPLC). For this purpose, the aboveground parts of the pine seedlings were homogenized by grinding in liquid nitrogen. One milliliter of 5% trichloroacetic acid (TCA) was added to 200 mg of the homogenized tissues. After vigorous mixing, the samples were incubated for 15 min at 75 °C. Samples were then centrifuged at

16,000 × *g* for 10 min. Supernatants were retained, and the pellets were re-extracted in the same manner with 0.5 mL of 5% TCA. After centrifugation, the pellets were discarded and the supernatants were pooled with those obtained after the first extraction step. Subsequently, the TCA extracts were extracted with portions of ethyl acetate in volumes of 500 µL (first portion) and 300 µL (second and third portion). All three portions of ethyl acetate were pooled and then subjected to drying in a vacuum centrifuge at 60 °C. The residues were dissolved in 200 µL of methanol, and the resultant solutions were analyzed using RP-HPLC. Chromatographic separations were carried out at room temperature using a HPLC system composed of a pump (model 1525, Waters), an autosampler (model 717plus, Waters), a fluorescence detector (model 474, Waters, detection was carried out at λ_{ex} = 310 nm/λ_{em} = 410 nm), C8 column (Symmetry 4.6 × 150 mm, 5 µm, Waters) connected to a C8 guard column (Symmetry 3.9 × 20 mm, 5 µm, Waters), and a personal computer with Breeze software (Waters). We used a 60 mM phosphate buffer at a pH 6.8 and acetonitrile with H_2O (4:1) as eluents. Chromatographic separations were carried out at a flow rate of 1 mL/min using a gradient elution. The sample injection volume was 25 µL. The concentration of salicylic acid in the samples was determined from the area of the corresponding peak on the basis of the prepared calibration curve.

2.5. Statistical Analyses

Before significance tests, the normality of the data distribution was verified by the Shapiro-Wilk test and the equality of variances was assessed by the Levene test. For many parameters (shoot length, root collar diameter, dry mass of aboveground parts and roots, and number of seedlings/m of row), the data distribution did not differ significantly from the normal distribution, and the variances of the compared variants were homogeneous, so the significance of differences was assessed by a one-way analysis of variance (ANOVA) and a *post hoc* Fisher's least significance difference (LSD) test. For the salicylic acid concentration, a two-way ANOVA was used (seedling treatment and date of assessment). For the remaining traits (number of lateral shoots, root length, number of root tips), nonparametric tests were used (non-parametric ANOVA,*i.e.*, Kruskal-Wallis test).

3. RESULTS

During the first inventory, an average of 103 seedlings/m of row were recorded in the variant treated with fungicides. In the other variants, seedling numbers were significantly smaller: control 78, chitosan 77 (p = 0.0015, Figure 1a). Relative to the number of sown seeds, seedling emergence reached 57% in the variant treated with fungicides and 48% in the remaining variants.

During the second inventory, an average 72 seedlings/m of row survived after fungicide applications, which corresponds to 40% of the sown seeds. In the variant with chitosan, 52 seedlings (29%) were recorded, compared to 37 seedlings (20%) in the control (Figure 1b). Differences between the variants were significant ($p < 0.0001$). Seedling emergence was affected most strongly by parasitic damping-off, and its symptoms were observed in all variants of the experiment. Symptoms of*Lophodermium* needle cast were not noticeable in any of the experimental variants.

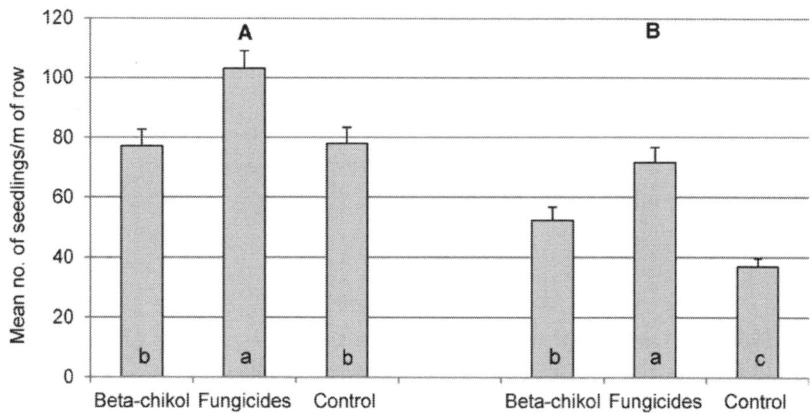

Figure 1. Mean number of Scots pine seedlings in 1-meter segments of the seed row (**A**) after seed germination and (**B**) approximately 7 weeks after germination in variants with Beta-chikol, fungicides, and no protection (control). Means marked with different letters are significantly different ($p < 0.05$, Fisher's LSD test). Error bars denote standard deviation.

All the analyzed biometric parameters of pine were affected similarly. The highest significant values of individual growth parameters were in the group treated with chitosan: shoot length 9.3 cm ($p < 0.0001$), root collar diameter 2.23 mm ($p < 0.0001$), number of lateral shoots 3.5 ($p < 0.0001$), aboveground dry weight 0.737 g ($p < 0.0001$), root dry weight 0.319 g ($p < 0.0001$), root length 332.7 cm ($p < 0.0001$), and number of root tips 796 ($p < 0.0001$). Pine seedlings protected with fungicides and unprotected seedlings (control) had lower values of these parameters, which did not differ significantly (Table 1).

The salicylic acid content in the seedlings five days after treatment ranged from 81.03 ng/g fresh weight (in one sample from the group treated with fungicides) to 113.56 ng/g fresh weight (in one of the control samples). On the date of the first assessment, the salicylic acid content of seedlings did not differ significantly. However, 12 days after treatment, the salicylic acid content was the highest in control pines: on average 188.26 ng/g fresh weight (treatment type $p < 0.0001$, date of assessment $p < 0.0001$, treatment type × date of assessment $p < 0.0001$) (Figure 2).

4. DISCUSSION

The forest nursery of the Spała Forest District has produced seedlings of forest trees for 50 years. The risk of parasitic damping-off (caused by fungi) is high in the nursery, which is confirmed by the results of our study. In the variant treated with fungicides, healthy seedlings developed from only 40% of sown seeds, while in the other variants, the corresponding values were even lower.

Chitosan applied in the form of Beta-chikol limited the rate of infection of pine seedlings with parasitic damping-off but was less effective than fungicides. The major reason for the low effectiveness was the lack of effect of seed treatment. Seedling emergence after seed priming with Beta-chikol was very similar to that of the control variant (no seed treatment).

Many studies indicate that the treatment of seeds of various herbaceous plants with chitosan increases the number of germinated seeds and decreases the number of infected seedlings [20,33], but the effectiveness of chitosan as a seed treatment depends on its molecular weight and concentration, the pH of the solution and the duration of seed soaking or number of layers of seed treatment powder [19,33]. Our results indicating a lack of protective effect of the Beta-chikol used for seed priming could be due to its application according to the manufacturer's instructions without taking into account the factors mentioned above. Unfortunately, published data do not allow us to determine unambiguously which values of individual factors are optimal. The varied effect of chitosan on seed germination may also be associated with local environmental conditions, e.g., the temperature and moisture of the soil surrounding the treated seeds [30,34].

Our results may be because the effectiveness of chitosan (applied as Biochikol 020 PC) as a forest seed treatment depends on the tree species, the fungal species causing the damping-off, and the concentration of the working solution. In a pot experiment, a seed treatment with 1% and 2% Biochikol proved to be ineffective at protecting against *Rhizoctonia solani* [35].

The effectiveness of pine protection against parasitic damping-off by foliar spraying with Beta-chikol in this study was comparable with that of chemical protection. Similar results of experiments with woody and herbaceous plants have been reported by many authors, e.g., [30,35]. The lack of symptoms of *Lophodermium* needle cast made it impossible to check the effectiveness of the tested product in protection against this disease.

Sathiyabama and Balasubramanian [12] showed that foliar application of chitosan on *Arachis hypogaea* caused a nearly four-fold increase in endogenous salicylic acid 8 days after treatment and more than a seven-fold increase after 12 days. Salicylic acid is a signal molecule that induces a cascade of signals that trigger plant defense reactions [23]. In our experiment, a significant increase in salicylic acid content was observed only in unprotected seedlings during the second assessment (12 July). Such a result indicates that the plant initiated natural defense responses induced by infection with the pathogen [23].

Table 1. Biometric parameters (mean and standard deviation in brackets) of Scots pine (*Pinus sylvestris*) protected with Beta-chikol, fungicides, and of the unprotected control.

Treatment	Shoot Length (cm)	Root Collar Diameter (mm)	Aboveground Dry Mass (g)	Root Dry Mass (g)	No. of Lateral Shoots	Root Length (cm)	No. of Root Tips
Beta-chikol	9.3 a	2.2 a	0.737 a	0.318 a	3.9 a	332.7 a	795 a
	(1.9)	(0.4)	(0.287)	(0.138)	(1.7)	(107.5)	(222)
Fungicides	8.1 b	1.8 b	0.515 b	0.183 b	3.2 b	178.4 b	424 b
	(1.8)	(0.4)	(0.261)	(0.091)	(2.5)	(85.1)	(182)
Control	7.6 b	1.6 b	0.455 b	0.169 b	2.3 b	192.0 b	486 b
	(1.5)	(0.4)	(0.208)	(0.078)	(2.3)	(89.7)	(175)

Means marked with different letters within columns are significantly different ($p < 0.05$).

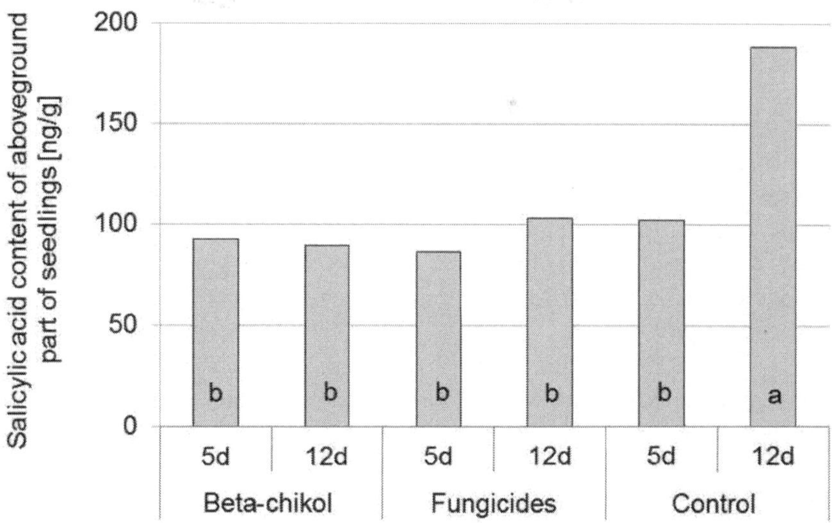

Figure 2. Salicylic acid content of Scots pine seedlings 5 and 12 days after chitosan treatment in variants with Beta-chikol, fungicides, and no protection (control). Explanations are as in Figure 1.

Chitosan application, in addition to causing a rise in the salicylic acid concentration, may trigger many other plant defense responses [8], which could also be the case in our experiment. The effective protection of seedlings against parasitic damping-off by foliar application of Beta-chikol, without any simultaneous increase in the salicylic acid content of the plants, suggests that some other defense mechanism(s) could be initiated, unrelated to the plant reaction. Patkowska *et al.*[36] reported that legume seed treatment and spraying with a solution containing chitosan (Biochikol 020 PC) resulted in microbiological changes in the rhizosphere. The treatment resulted in a 30% lower number of isolates of pathogenic fungi, including those causing parasitic damping-off, and higher numbers of isolates of bacteria of the genera *Bacillus* and*Pseudomonas*, as well as fungi of the genera *Aspergillus*, *Gliocladium*, and *Trichoderma*. The stimulation of growth of these groups of soil microorganisms results from their ability to synthesize chitinases, *i.e.*, the enzymes that determine their resistance to the influence of chitosan [8,21]. Simultaneously, the high chitinolytic activity makes it possible to use the above-mentioned genera of bacteria and fungi for plant production, as plant-growth-promoting rhizobacteria (PGPR) and plant-growth-promoting fungi (PGPF). These bacteria and fungi induce plant resistance to diseases, abiotic stresses, stimulate plant growth, and indirectly affect pathogenic fungi [37,38].

The effectiveness of chitosan in plant protection in this study could also result from a direct effect of the Beta-chikol on the fungi and oomycetes causing the parasitic damping-off, reflected in the inhibition of hyphal growth and spore germination [17,39]. Another mechanism could be the ability of chitosan molecules to bind mycotoxins secreted by facultative pathogens, such as those causing damping-off, to enable them to colonize plant tissues [40].

Chitosan applied as Beta-chikol resulted in the intensive stimulation of pine growth. Chitosan applies to all the analyzed growth parameters of both the roots and aboveground parts. A similar stimulation of growth was observed in many other studies [19,41,42,43]. The mechanism of this phenomenon has not been fully explained so far. The stimulation can result from a direct effect of chitosan uptake by roots and use in metabolic processes by the plant as an additional source of nutrients [42]. Chitosan is a rich source of nitrogen and calcium and contains also micronutrients, including copper, zinc, and iron [40,44]. Better availability of nutrients to the plants may also result from the chelating properties of chitosan [45]. Moreover, the plant growth stimulation may be an effect of the microbial changes in the soil described above. The increased abundance of PGPR and PGPF leads to better growth of plants [37,38].

There probably exists a still unknown mechanism of plant growth stimulation as a result of chitosan application because this phenomenon was also observed in experiments with very low concentrations of chitosan, making it impossible to use them as a source of nitrogen and carbohydrates. Moreover, plant growth was also stimulated by chitosan in experiments performed in aseptic conditions, where the effects of PGPR and PGPF were impossible [21].

5. CONCLUSIONS

Chitosan is an organic, environmentally friendly, biodegradable product that is non-toxic to higher organisms. Its effectiveness in the protection of Scots pine seedlings against parasitic damping-off by foliar spraying is comparable with that of a fungicide treatment. It can be used in forest nurseries as an alternative to chemical methods and as a product stimulating seedling growth. Modern forest management in the European Union seeks environmentally friendly technological solutions; thus, chitosan should find its place in the Integrated Pest Management (IPM) systems of forest nurseries.

ACKNOWLEDGMENTS

We thank Przemysław Grabowski (Manager of Spała Forest District), Tadeusz Długołęcki (Deputy Manager), and Joanna Choromańska for allowing us to perform this experiment and Waldemar Socha (nurseryman) for help during this study. The editorial help of Sylwia Ufnalska is also gratefully acknowledged.

AUTHOR CONTRIBUTIONS

Marta Aleksandrowicz-Trzcińska: idea, laboratory work, data interpretation and writing; Anna Bogusiewicz: filed and laboratory work; Michał Szkop: measured the salicylic acid concentration; Stanisław Drozdowski: performed the statistical analysis.

REFERENCES

1. Thakur, V.K.; Kesster, M.R. Self-healing polymer nanocomposite materials: A review. *Polymer* **2015**, *69*, 369–383.

2. Thakur, V.K.; Thakur, M.K. Recent advances in green hydrogels from lignin: A review. *Int. J. Biol. Macromol.* **2015**, *72*, 834–847.

3. Pappu, A.; Patil, V.; Jain, S.; Mahindrakar, A.; Haque, R.; Thakur, V.K. Advances in industrial prospective of cellulosic macromolecules enriched banana biofibre resources: A review. *Int. J. Biol. Macromol.* **2015**, *79*, 449–458.

4. Thakur, V.K.; Thakur, M.K.; Gupta, R.K. Review: Raw natural fiber-based polymer composites. *Int. J. Polym. Anal. Charact.* **2014**, *19*, 256–271.

5. Thakur, V.K.; Thakur, M.K. Processing and characterization of natural cellulose fibers/thermoset polymer composites.*Carbohydr. Polym.* **2014**, *109*, 102–117.

6. Thakur, V.K.; Thakur, M.K. Recent advances in graft copolymerization and applications of chitosan: A review. *ACS Sustain. Chem. Eng.* **2014**, *2*, 2637–2652.

7. Chandler, D.; Bailey, A.S.; Tatchell, G.M.; Davidson, G.; Greaves, J.; Grant, W.P. The development, regulation and use of biopesticides for integrated pest management. *Philos. Trans. R. Soc. B* **2011**, *366*, 1987–1998.

8. El Hadrami, A.; Adam, L.R.; El Hadrami, I.; Daayf, F. Chitosan in plant protection. *Mar. Drugs* **2010**, *8*, 968–987.

9. Raafat, D.; Sahl, H.-G. Chitosan and its antimicrobial potential—A critical literature survey. *Microb. Biotechnol.* **2009**, *2*, 186–201.

10. Manjunatha, G.; Niranjan Raj, S.; Prashanth, G.N.; Deepak, S.; Amruthesh, K.N.; Shetty, H.S. Nitric oxide is involved in chitosan-induced systemic resistance in pearl millet against downy mildew disease. *Pest Manag. Sci.* **2009**, *65*, 737–743.

11. Zuppini, A.; Baldan, B.; Millioni, R.; Favaron, F.; Navazio, L.; Mariani, P. Chitosan induces Ca^{2+}-mediated programmed cell death in soybean cells. *New Phytol.* **2003**, *161*, 557–568.

12. Sathiyabama, M.; Balasubramanian, R. Chitosan induce resistance components in *Arachis hypogaea* against leaf rust caused by *Puccinia arachidis* Speg. *Crop Prot.* **1998**, *17*, 307–313.

13. Reddy, M.V.B.; Arul, J.; Angers, P.; Couture, L. Chitosan treatment of wheat seeds induces resistance to *Fusarium graminearun* and improves seed quality. *J. Agric. Food Chem.* **1999**, *47*, 1208–1216.

14. Fajardo, J.E.; Waniska, R.D.; Cuero, R.G.; Pettit, R.E. Phenolic compounds in peanut seeds: Enhanced elicitation by chitosan and effects on growth and aflatoxin B_1 production by *Aspergillus flavus. Food Biotechnol.* **1995**, *9*, 59–78.

15. Pospieszny, H.; Chirkov, S.; Atabekov, J. Induction of antiviral resistance in plants by chitosan. *Plant Sci.* **1991**, *79*, 63–68.

16. Li, B.; Liu, B.; Shan, C.; Ibrahim, M.; Lou, Y.; Wang, Y.; Xie, G.; Li, H.; Sun, G. Antibacterial activity of two chitosan solutions and their effect on rice bacterial leaf blight and leaf streak. *Pest Manag. Sci.* **2013**, *69*, 312–320.

17. Laflamme, P.; Benhamou, N.; Bussières, G.; Dessureault, M. Differential effects of chitosan on root rot fungal pathogens in forest nurseries. *Can. J. Bot.* **2000**, *77*, 1460–1468.

18. Hernández-Lauzardo, A.N.; Vega-Pérez, J.; Velázquez-Del Valle, M.G.; Sánchez, N.S.; Peña, A.; Guerra-Sánchez, G. Changes in the functionality of plasma membrane of *Rhizopus stolonifer* by addition of chitosan. *J. Phytopathol.* **2011**,*159*, 563–568.

19. Cho, M.H.; No, H.K.; Prinyawiwatkul, W. Chitosan treatments affect growth and selected quality of sunflower sprouts. *J. Food Sci.* **2008**, *73*, 70–77.

20. Manjunatha, G.; Roopa, K.S.; Prashanth, G.N.; Shetty, H.S. Chitosan enhances disease resistance in pearl millet against downy mildew caused by *Sclerospora graminicola* and defence-related enzyme activation. *Pest Manag. Sci.* **2008**, *64*, 1250–1257.

21. Sharp, R.G. A review of the applications of chitin and its derivatives in agriculture to modify plant-microbial interactions and improve crop yields. *Agronomy* **2013**, *3*, 757–793.

22. Raskin, I.; Skubatz, H.; Tang, W.; Meeuse, B.J.D. Salicylic acid levels in thermogenic and non-thermogenic plants. *Ann. Bot.* **1990**, *66*, 369–373.

23. Malamy, J.; Carr, J.P.; Klessig, D.F.; Raskin, I. Salicylic acid: A likely endogenous signal in the resistance response of tobacco to viral infection. *Science* **1990**, *250*, 1002–1004.

24. Raskin, I. Role of salicylic acid in plants. *Annu. Rev. Plant Physiol. Plant Mol. Biol.* **1992**, *43*, 439–463.

25. Janda, T.; Szalai, G.; Tari, I.; Páldi, E. Hydroponic treatment with salicylic acid decreases the effects of chilling injury in maize (*Zea mays* L.) plants. *Planta* **1999**, *208*, 175–180.

26. Kaydan, D.; Yagmur, M.; Okut, N. Effects of salicylic acid on the growth and some physiological characters in salt stressed wheat (*Triticum aestivum* L.). *Tarim Bilim. Derg.* **2007**, *13*, 114–119.

27. Żurek, A.; Grzyś, E. Wpływ kwasu salicylowego na parametry fizjologiczne siewek ogórka traktowanych herbicydem Reglone. *Post. Ochr. Roślin* **2007**, *47*, 325–329, (In Polish with English summary).

28. Matysiak, K.; Adamczewski, K. Regulatory wzrostu i rozwoju roślin—Kierunki badań w Polsce i na świecie. *Post. Ochr. Roślin* **2009**, *49*, 1810–1816, (In Polish with English summary).

29. Handro, W.; Mello, C.M.; Manzano, M.A.; Floh, E.I.S. Enhancement of stem elongation and flower bud regeneration by salicylic acid. *Rev. Bras. Fisiol. Veg.* **1997**, *9*, 139–142.

30. Hadwiger, L.A. Multiple effects of chitosan on plant system: Solid science or hype. *Plant Sci.* **2013**, *208*, 42–49.

31. Forest Management Plan 2012–2021, Spała Forest District, Spała, Poland, Unpublished work. 2012. (In Polish).

32. Głowacka, B. *Środki Ochrony Roślin Oraz Produkty do Rozkładu Pni Drzew Leśnych Zalecane do Stosowania w Leśnictwie w Roku 2013*; IBL: Sękocin Stary, Poland, 2012; pp. 1–72.

33. Ziani, K.; Ursúa, B.; Maté, J.I. Application of bioactive coatings based on chitosan for artichoke seed protection. *Crop Prot.* **2010**, *29*, 853–859.

34. Guan, Y.J.; Hu, J.; Wang, X.J.; Shao, C.X. Seed priming with chitosan improves miaze germination and seedling growth in relation to physiological changes under low temperature stress. *J. Zhejiang Univ.—Sci. B* **2009**, *10*, 427–433.

35. Duda, B.; Oszako, T.; Piwnicki, J. Possibilities of chitosan use in forestry. *Bull. Pol. Acad. Sci. Biol. Sci.* **2003**, *51*, 213–220.

36. Patkowska, E.; Pięta, D.; Pastucha, A. *The Effect of Biochikol 020 PC on Microorganism Communities in the Rhizosphere of Fabaceae Plants*; Polish Chitin Society: Łódź, Poland, 2006; pp. 171–178.

37. Harman, G.E. Multifunctional fungal plant symbionts: New tools to enhance plant growth and productivity. *New Phytol.* **2011**, *189*, 647–649.

38. Ramamoorthy, V.; Viswanathan, R.; Raguchander, T.; Prakasam, V.; Samiyappan, R. Induction of systemic resistance by plant growth promoting rhizobacteria in crop plants against pests and diseases. *Crop Prot.* **2001**, *20*, 1–11.

39. Palma-Guerrero, J.; Jansson, H.-B.; Salinas, J.; Lopez-Llorca, L.V. Effect of chitosan on hyphal growth and spore germination of plant pathogenic and biocontrol fungi. *J. Appl. Microbiol.* **2008**, *104*, 541–553.

40. Bornet, A.; Teissedre, P.L. Chitosan, chitin-glucan and chitin effects on minerals (iron, lead, cadmium) and organic (ochratoxin A) contaminants in wines. *Eur. Food Res. Technol.* **2008**, *226*, 681–689.

41. Aleksandrowicz-Trzcińska, M.; Hamera-Dzierżanowska, A.; Żybura, H.; Drozdowski, S. Wpływ mykoryzacji i chitozanu na wzrost sosny zwyczajnej (*Pinus sylvestris* L.) w szkółce i na uprawie. *Sylwan* **2013**, *157*, 899–908, (In Polish with English summary).

42. Ohta, K.; Morishita, S.; Suda, K.; Kobayashi, N.; Hosoki, T. Effects of chitosan soil mixture treatment in the seedling stage on the growth and flowering of several ornamental plants. *J. Jap. Soc. Hortic. Sci.* **2004**, *73*, 66–68.

43. Ohta, K.; Taniguchi, A.; Konishi, N.; Hosoki, T. Chitosan treatment affects plant growth and flower quality in *Eustoma grandiflorum*. *HortScience* **1999**, *34*, 233–234.

44. Ramírez, M.Á.; Rodriguez, A.T.; Alfonso, L.; Peniche, C. Chitin and its derivatives as biopolymers with potential agricultural applications. *Biotecnol. Apl.* **2010**, *27*, 270–276.

45. Ravi Kumar, M.N.V. A review of chitin and chitosan applications. *React. Funct. Polym.* **2000**, *46*, 1–27.

Chapter 8

Designing Strategies for Epidemic Control in a Tree Nursery: the Case of Ash Dieback in the UK

Vasthi Alonso Chavez [1,*], Stephen Parnell [1,2] and Frank van den Bosch [1]

[1]Department of Computational and Systems Biology, Rothamsted Research, Harpenden, AL5 2JQ, UK

[2]School of Environment and Life Sciences, University of Salford, Manchester, M5 4WT, UK

ABSTRACT

Ash dieback is a fungal disease (causal agent *Hymenoscyphus fraxineus*) infecting Common ash (*Fraxinus excelsior*) throughout temperate Europe. The disease was first discovered in the UK in 2012 in a nursery in Southern England, in plants which had been imported from the Netherlands. After sampling other recently planted sites across England, more infected trees were found. Tree trade from outside and across the UK may have facilitated the spread of invasive diseases which threaten the sustainability of forestry business, ecological niches and amenity landscapes. Detecting a disease in a nursery at an early stage and knowing how likely it is for the disease to have spread further in the plant trade network, can help control an epidemic. Here, we test two simple sampling rules that 1) inform monitoring strategies to detect a disease at an early stage, and 2) inform the decision of tracking forward the disease after its detection. We apply these expressions to the case of ash dieback in the UK and test them in different scenarios after disease introduction. Our results are useful to inform policy makers' decisions on monitoring for the control and spread of tree diseases through the nursery trade.

KEYWORDS

ash dieback; sampling; selling; epidemic; incidence; monitoring; nursery

1. INTRODUCTION

Common ash is an iconic tree across temperate Europe, being widespread in broadleaved woodlands. It has a natural range that extends from Southern Scandinavia to Northern Spain and the Balkans, and from Ireland in the west to continental Russia in the east [1]. Common ash is usually found in mixed woodland communities with trees, such as beech, oak, alder, and sycamore [1,2], and accounts for 14% of the total broadleaved standing volume in the UK, making it the fourth most common broadleaved tree [3]. Ash is an important species for the functioning and conservation of forest ecosystems throughout Europe and the UK and is also an economically important species, having a valuable and high in demand timber because of its physical properties [2].

Ash dieback (causal agent *Hymenoscyphus fraxineus*) is a fungal disease which affects several species of the *Fraxinus* genus including European or common ash (*Fraxinus excelsior*), which is the most affected [4]. It was first observed in Poland in the early 1990s, but the disease has continued its spread across Europe, reaching a large proportion of countries including the UK with the first ash dieback observation in a tree nursery in Buckinghamshire in February 2012 [5].

Although ash dieback had been observed since the early 1990s, the anamorph (asexual stage) of the causal organism (*Chalara fraxinea*) was named only in 2006 [6]. By 2009, *C. fraxinea* was wrongly considered to be the anamorph of*Hymenoscyphus albidus*, a species considered to be non-pathogenic, native and widespread in Europe [7]. In 2010, it was found that the sexual stage of *C. fraxinea* was *Hymenoscyphus pseudoalbidus* [8], morphologically similar to, but genetically distinct from *H. albidus*. It was only in 2014 that the correct scientific name *Hymenoscyphus fraxineus* was given for the fungus causing ash dieback in Europe [9]. This meant that before 2012, the pathogen was not designated as an organism of statutory concern under the EU Plant Health Directive. Thus, ash trade was allowed to continue to spread in the UK and other European countries, despite concerns and attempts to ban ash imports into the UK from the Horticultural Trades Association (HTA) in 2009 [10] due to the spread of the disease across Europe.

Once the disease was detected in the UK, a rapid assessment of the disease was undertaken by Forest Research to determine the status of the pathogen [11] followed by a pest risk analysis [5] and a Chalara (ash dieback) management plan [12]. These assessments determined that the disease was not only found in several nurseries but was also discovered in the wider environment. "Tracing forward" from infected nurseries in order to determine sites with recently planted trees was one of the measures taken to reduce the rate of epidemic spread [12]. Finding the disease in nurseries also triggered the assessment of the disease in the wider environment, allowing the detection and the development of a plan to slow the disease spread [12].

It has become clear that one of the possible paths for ash dieback spread is through trade. Nurseries are strategic points for the distribution of trees and the potential spread of disease. Therefore, understanding the processes of disease-monitoring within a nursery, and plant sales from a nursery can help developing sampling and tracking strategies to slow and control the spread of diseases introduced in nurseries via trade.

In the UK, plant health inspections are undertaken by the Department for Environment, Food and Rural Affairs (Defra) and the Forestry Commission with the aim of keeping non-native species that may affect crops, trees and wild plants out of the country [13]. In the case of the hard nursery stock sector, the Plant Health and Seeds Inspectorate (PHSI) sets policy and enforces controls and restrictions on the import, movement and keeping of certain plants, plant pests and other materials such as soil.

Developing tools that can help inspectors and business owners detect and control a disease before it becomes widespread are needed. In the UK nursery inspections usually happen from July–September since this time of the year is when disease expression is clearer. Official inspections for nurseries can happen between two and four times per year, depending on the business's risk to plant health [14].

If a diseased tree shows no symptoms during inspection, the disease may continue spreading to other hosts or even moved to other areas if sold. Therefore, here we present a method that accounts for disease asymptomatic periods that play an important role for the epidemic development.

We describe a simple monitoring method for a single nursery in order to (1) detect a disease at an early stage, considering that the disease has an asymptomatic period, and (2) determine the need of tracking forward the disease depending on the probability of having sold or moved diseased plants to other places. We apply this method to the case of ash dieback and consider different control scenarios to understand the best strategies to slow the disease spread.

2. MODEL AND METHODS

In this section, we describe the monitoring method developed in two parts. The first part of the method determines sampling strategies that help detecting ash dieback in a nursery at an early epidemic stage. The second helps understanding what strategies would have been needed to reduce the probability that infected plants were moved to other points in the tree trade network, such as recently planted sites, when an epidemic is first discovered.

2.1. Epidemic Dynamics

We assume that the disease arrives to a nursery at time t_0 affecting the tree population with an initial incidence (q_0) (the initial proportion of affected trees), assumed to be small in comparison with the total number of trees. If we aim to detect the epidemic at an early stage, we can assume that the epidemic follows logistic growth so that the epidemic incidence over time is given by:

$$q(t) = \frac{q_0 \exp\big(r(t - t_0 + \lambda)\big)}{1 - q_0 + q_0 \exp\big(r(t - t_0 + \lambda)\big)} \tag{1}$$

where the incidence $q(t)$ is the proportion of individuals infected (symptomatic and asymptomatic) at time t from the total host population [15], r is the epidemic growth rate, λ the epidemic asymptomatic period and q_0 the initial incidence at the time t_0, when the disease arrives to the nursery.

2.2. Monitoring Plan in the Nursery

We assume that a monitoring scheme for nurseries is already in place before the epidemic starts. In this monitoring scheme, N hosts are randomly sampled every fixed period of time Δt. We also assume that only symptomatic hosts can be identified through visual inspection. We then assume that the probability of detecting at least one infected individual at $t_1$given that epidemic started at t_0 follows a binomial distribution and it is given by:

$$P(t_1|t_0) = 1 - \left(1 - q_s(t_1)\right)^N. \tag{2}$$

The probability of not being able to detect any infected hosts throughout all sampling rounds previous to t_1 is:

$$\prod_{i=1}^{k}\left(1 - q_s(t_1 - i\Delta t)\right)^N, \tag{3}$$

where k is the total number of sampling rounds between t_0 and t_1. Using these assumptions Alonso Chavez *et al.* [16] found a rule of thumb to determine the expected incidence of the disease at first detection, q^*:

$$E(q^*) \approx \frac{r\,\Delta t}{N} e^{r\lambda}, \tag{4}$$

where λ is the disease's asymptomatic period [16]. This approximation is a simple method to determine a sampling strategy that enables early detection of a disease.

Equation (4) has been tested in practical situations and has been found to give accurate predictions under conditions where the asymptomatic period is absent (or where asymptomatic infection can be detected by the detection method, e.g., PCR) [17,18]. It says that the expected discovery incidence depends on the epidemic growth rate, the sampling rate and the length of the asymptomatic period. If we know the epidemic growth and the asymptomatic period of a disease, it is possible to design a sampling strategy to detect a disease below a designated expected incidence $E(q^*)$.

2.3. Selling Trees from the Nursery

When a nursery is infected (either by buying one or more infected trees or by inoculum being blown into the nursery) an epidemic will develop and there is a risk that infected trees are sold onwards before the epidemic is detected.

To understand how tree sales affect the probability of having moved or sold infected trees from one place to another, we assume that a nursery has a strategy where M trees are sold every $\Delta\tau$ period of time. We overlay this strategy on the monitoring program described in Section 2.2. We then calculate the probability that no infected hosts were sold during any selling round previous to, and including t_2; the last selling time before disease detection at t_1.

The probability that no infected hosts were sold at t_2 given that epidemic started at t_0 and that disease was detected at $t_1 > t_2$ is given by:

$$= \left[1-(1-q(t_1))^N\right]\cdot\prod_{i=1}^{k}(1-q(t_1-i\Delta t))^N\cdot\prod_{j=0}^{l}(1-q(t_2-j\Delta\tau+\lambda))^M \tag{5}$$

where $q(t_1)$ is the disease incidence at discovery time, $q(t_2)$ is the incidence at the last selling time before discovery, λ is the asymptomatic period, N and M are the sampling and selling host numbers respectively, Δt and $\Delta\tau$ are the sampling and selling interval periods, k is the total number of sampling rounds between t_0 and t_1, and l is the total number of selling rounds between t_0 and t_2.

If we sum all possible discovery times t_1 Equation (5) can be simplified to find the approximate solution [16]:

$$P(\text{no infected sales before detection}) \approx \frac{\frac{N}{\Delta t}}{\frac{N}{\Delta t}+\frac{M}{\Delta \tau}e^{r\lambda}} \tag{6}$$

Equation (6) notes that the probability of having sold infected hosts depends on the sampling rate, the selling rate, the epidemic growth rate and the asymptomatic period of the disease. Notice that if we are looking for the probability that no infected hosts were sold to be large, the sampling rate has to be much larger that the term related to the selling rate.

2.4. Asymptomatic Period for Ash Dieback

The incubation period of *H. fraxineus* after ascospore inoculation has been found to be 10–14 days, with hosts showing initial symptoms, such as necrotic lesions in leaf tissue after 2 weeks of ascospore exposure [19,20]. However, visible symptoms at plant scale can appear after several months and up to a year. It has been observed that apparently disease-free seedlings that show no symptoms during the late autumn, show ash dieback symptoms by the beginning of the following year's spring [4]. Moreover, if a seedling is infected in the spring or summer of year 1, visible symptoms may either appear during the late summer or autumn of the same year, or even in the spring of the next year [5]. This means that an infected host can be asymptomatic for up to one year. This, up to one-year-long asymptomatic period, can be identified with shoot symptoms (bark lesions) that usually arise during the dormancy period of the host, after leaf shed, when symptoms cannot be easily identified through visual inspection. Under these premises, we assume that the asymptomatic period of ash dieback is between 14 days–1 year since sampling is usually done at a plant scale.

2.5. Epidemic Growth Rate for Ash Dieback

Studies in experimental plots recorded the temporal development of ash dieback. In south-western Germany an ash field trial was established in 2005 [21]. Following a temporal development of the disease, 3-year-old ash trees were surveyed in replicated blocks or 36 plants. The development of proportion of classes of disease severity over time was recorded. We assume that the incidence in 2005, when the field trial was established is $q = 0$, while, for 2006, we assume very low incidence ($q = 0.01$). For the years 2007–2012 we used the data shown in [21] and recorded the percentage of total infected trees belonging to all severity classes. From this information, we calculated the epidemic growth rate of the disease using the least-squared method, assuming that the epidemic followed logistic growth. The epidemic growth rate obtained for the period 2005–2012 was $r = 0.0026/day$ (Figure 1).

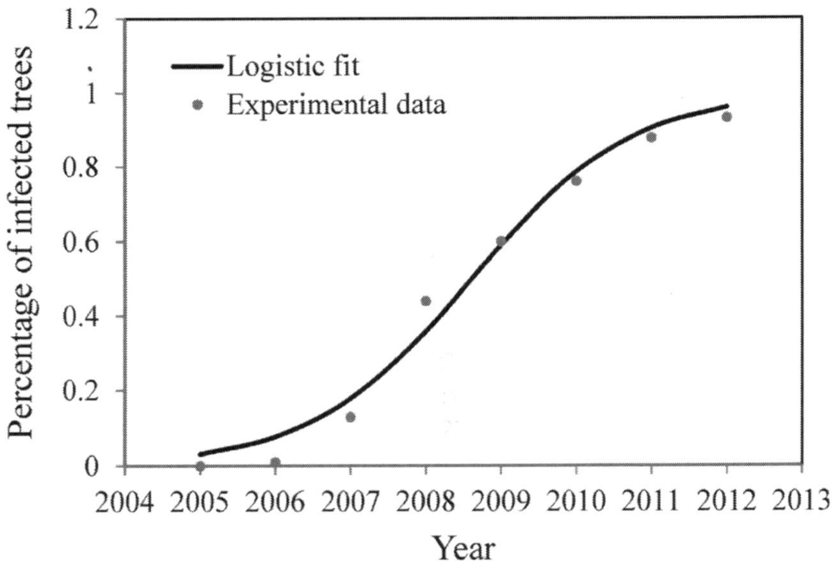

Figure 1. Ash dieback incidence is fitted to a logistic curve using least-squares method. The dots denote the experimental data recorded by Enderle *et al*. [21]. The solid line corresponds to the disease fitted curve.

The logistic epidemic growth assumed to obtain the epidemic growth rate is a time-scale approximation that does not take into account seasonality of the disease spread, which is important for symptoms appearance and disease development [22]. Nonetheless, it has been shown that in cases such as *Citrus canker* where epidemic can increase in bursts whenever there is a storm, this approximation is robust despite this temporal variability in epidemic spread [18,23].

3. RESULTS AND DISCUSSION

In this section we apply the sampling expressions shown in Equations (4) and (6) to ash dieback for the epidemic growth rate and asymptomatic period shown in Section 2.2 and Section 2.3. We determine sampling and selling strategies that would help detecting the disease at low incidence and that would prevent its spread to other points in the plant trade-network.

We assume that the growth rate of the epidemic is $r = 0.0026$ and that the asymptomatic period is approximately 150 days (five months). We take this asymptomatic period based on the information from [4] where it is observed that it is common to see that after a host is infected during late summer or early autumn, it starts showing symptoms in early spring.

Before the discovery of ash dieback in the UK in 2012, a large nursery could produce between 500,000–1,000,000 trees per year, selling up to ~300,000 trees in one month. Inspections would occur between two and four times per year, but as common ash was not a regulated species, only casual inspections for diseases in ash would take place.

Figure 2A shows how the expected incidence at discovery depends on the sampling size N for different monitoring periods (Δt = 60, 90, 120, 180 days). The expected incidence decreases as the sampling size increases and as the sampling interval decreases, the detected expected incidence decreases as well. Figure 2A also shows that in order to detect the disease at an expected incidence below 1%, an inspector would have to sample between 10 and 40 hosts, depending on the sampling interval. For example, if an inspector were to monitor a nursery every three months (solid line), sampling at least 16 hosts will allow detecting the disease at an expected incidence of 1%.

In summary, Figure 2A shows that it is possible to design sampling strategies to detect ash dieback at low incidence. This means that if the already established monitoring process before 2012 had considered ash dieback as a threat before its discovery, the probability of detecting the disease at low incidence would have been a real possibility.

It is important to note that, an ideal monitoring situation would be one where the sampling strategy could detect the pathogen in the early asymptomatic growth phase, using, for example, a sensitive molecular diagnostic assay. In such a scenario, the asymptomatic period λ can be neglected. However, a common monitoring strategy is to perform visual inspection to detect symptomatic plants. Once a symptomatic host is discovered, PCR is used to confirm if the suspected disease exists. Including the asymptomatic period allows us to assess the real disease incidence when detected, considering symptomatic and asymptomatic hosts, when visual inspection is the principal monitoring tool. This reduces the probability of underestimating the disease incidence by neglecting the asymptomatic period of the disease. This also gives us a better estimate of the probability of disease spread to other points of the trade network, informing better strategies for disease tracking.

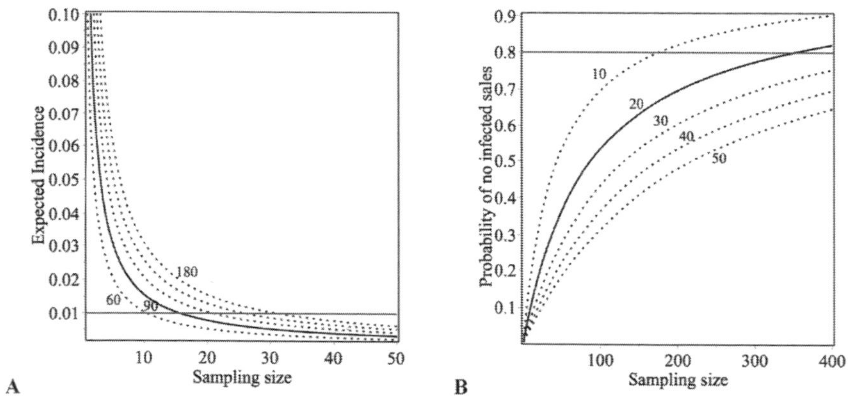

Figure 2. (**A**) The expected incidence for a number of sampling intervals (60, 90, 120, 150, and 180 days) depends on the sampling size for ash dieback. The solid line shows a typical sampling interval time between inspections (90 days). In order to detect the disease at incidences below 0.01 there are a number of sampling strategies which can be adopted. If the sampling size is over approximately 35 individuals, the detection incidence stays below 0.01. (**B**) The probability that no infected individuals were sold before disease detection, depends on the sampling size. For this example we take a sampling interval of Δt = 90 days and a selling interval of $\Delta\tau$ = 30 days. Solid and dotted lines show number of plants sold (10, 20, 30, 40, 50) dependent on the sampling size. To ascertain that the probability that no infected sales taking place before detection stays above 0.8, the sampling size has to be much larger than the selling size. The solid line shows that if the nursery sells 20 trees per month, a monitoring strategy that samples ~350 plants every three months is needed to maintain this probability.

Figure 2B shows how the probability that no infected trees were sold before detection depends on the sampling size *N* for different number of trees sold (M = 10, 20, 30, 40, 50). In this example we assume that the sampling interval is Δt = 90 and the selling interval is $\Delta \tau$ = 30. The probability that no infected trees were sold before detection increases as the number of sold trees decreases and as the sampling size increases. However, we note that the sampling rate has to be much larger compared with the selling rate in order to ensure that the probability that no infected plants were sold stays large enough *(p (no infected sales) > 0.8)*. For example, the solid line in Figure 2B shows that for a selling plan of 20 trees/30 days, a sampling strategy of 350 trees/90 days is needed to ensure that at least 80% of ash trees sold or moved are disease free. This sampling strategy does not stop the potential movement of infected material since the current selling dynamics of a big nursery involves the regular movement of large numbers of trees. This result implies that, even if ash dieback can be detected at low incidence, a tracking plan to identify recently planted sites with ash coming from nurseries were infected individuals are found, is necessary.

4. CONCLUSIONS

We conclude that quantitative estimates of the performance of a sampling plan for incoming diseases such as ash dieback are possible and can be used to allocate sampling efforts appropriately. The presented method provides a simple strategy to detect diseases at low incidence, and thus control the spread of potential incoming diseases. Here we show that with the monitoring methods tested in this manuscript, a sampling strategy involving the inspection of a low number of trees every three months, would have been sufficient to detect ash dieback (and other diseases with similar epidemic growth rates and asymptomatic periods) at incidences below 1% within an infected nursery. The monitoring method presented here has been tested against detailed simulation models and real data and has shown to be robust [18], providing great insight to develop monitoring strategies through a simple rule. However, it is important to note that increased accuracy can be obtained by including seasonality in symptom expression and transmission, and should be evaluated in future work.

We also show that given the UK existing monitoring strategies, the probability that once a disease has entered into a nursery, the disease is moved to other points is high. The expression given in Equation (8) shows that the sampling rate in a nursery has to be much bigger than the selling rate, in order to ensure the containment of the disease in the nursery. We show that in the case of ash dieback, the probability of not having sold or moved infected trees was higher than 0.8 only when the sampling rate was around 6 times the selling rate. Given that, in general the nursery businesses sell large volumes of trees, a well standardised and regulated regime able to declare any tree provenance and destination may help reducing tracking costs and disease spread.

The structure of the plant trade network in the UK and the EU in general, allows nursery owners and producers to trade with each other and with other type of businesses under certain well established rules. The methods presented here show that monitoring strategies can be designed in order to detect diseases at low incidence and track infected host from individual nurseries. However, the introduction and tracking of a disease into a single nursery are intrinsically linked to plant trade at larger scales. The methods presented here constitute a first step to develop broader strategies that can include a larger number and different types of stakeholders. Strategies that quantify the implications of monitoring strategies at national and international levels should be investigated further.

ACKNOWLEDGMENTS

Authors thank two anonymous reviewers for their helpful comments. Many thanks to J. Knight, D. McCann, D. Brown, K. Sacre, J. Dewhurst, M. Hommel, M. Robinson, B. Fraser, S. Ashworth, C. Carpenter, N. Darcy and P. Rochford for providing insight about the nursery industry in the UK. Rothamsted Research receives support from the Biotechnology and Biological Sciences Research Council (BBSRC).

AUTHOR CONTRIBUTIONS

Frank van den Bosch supervised the project; Vasthi Alonso Chavez, Stephen Parnell and Frank van den Bosch designed the study, developed the methodology and reviewed the manuscript. Vasthi Alonso Chavez performed the analysis and wrote the manuscript. All authors discussed the results and implications and commented on the manuscript at all stages.

REFERENCES

1. Beatty, G.E.; Brown, J.A.; Cassidy, E.M.; Finlay, C.M.V.; McKendrick, L.; Montgomery, W.I.; Reid, N.; Tosh, D.G.; Provan, J. Lack of genetic structure and evidence for long-distance dispersal in ash (*Fraxinus excelsior*) populations under threat from an emergent fungal pathogen: Implications for restorative planting. *Tree Genet. Genomes* **2015**, *11*, 1614–2942.

2. Pautasso, M.; Aas, G.; Queloz, V.; Holdenrieder, O. European ash (*Fraxinus excelsior*) dieback—A conservation biology challenge. *Biol. Conserv.* **2013**, *158*, 37–49.

3. National Forest Inventory, F.C. *NFI Preliminary Estimates of Quantities of Broadleaved Species in British Woodlands, with Special Focus on Ash*; Forestry Commission: Edinburgh, UK, 2013.

4. Kirisits, T.; Kritsch, P.; Kräutler, K.; Matlakova, M. Ash dieback associated with Hymenoscyphus pseudoalbidus in forest nurseries in Austria. *J. Agric. Ext. Rural Dev.* **2012**, *4*, 230–235.

5. Sansford, C.E. Pest Risk Analysis for Hymenoscyphus pseudoalbidus for the UK and the Republic of Ireland. *For. Comm.* **2013**, 1–128.

6. Kowalski, T. *Chalara fraxinea* sp. nov. associated with dieback of ash (*Fraxinus excelsior*) in Poland. *For. Pathol.* **2006**, *36*, 264–270.

7. Kowalski, T.; Holdenrieder, O. The teleomorph of Chalara fraxinea, the causal agent of ash dieback. *For. Pathol.* **2009**,*39*, 304–308.

8. Queloz, V.; Grünig, C.R.; Berndt, R.; Kowalski, T.; Sieber, T.N.; Holdenrieder, O. Cryptic speciation in Hymenoscyphus albidus. *For. Pathol.* **2011**, *41*, 133–142.

9. Baral, H.-O.; Queloz, V.; Hosoya, T. Hymenoscyphus fraxineus, the correct scientific name for the fungus causing ash dieback in Europe. *Int. Mycol. Assoc. Fungus* **2014**, *5*, 79–80.

10. Ash Dieback. Available online: http://www.the-hta.org.uk/ page.php?pageid=1001 (accessed on 22 September 2015).

11. Webber, J.; Hendry, S. *Rapid Assessment of the Need for A Detailed Pest Risk Analysis for Chalara Fraxinea*; Forest Research: Edinburgh, UK, 2012.

12. DEFRA. *Chalara Management Plan*; DEFRA: London, UK, 2013.

13. Williams, F.; Eschen, R.; Harris, A.; Djeddour, D.; Pratt, C.; Shaw, R.S.; Varia, S.; Lamontagne-Godwin, J.; Thomas, S.E.; Murphy, S.T. *The Economic Cost of Invasive Non-Native Species on Great Britain*; CABI: Wallingford, UK, 2010.

14. Issuing plant passports to trade plants in the EU. Available online: https://www.gov.uk/issuing-plant-passports-to-trade-plants-in-the-eu (accessed on 21 August 2015).

15. Madden, L.V.; Hughes, G.; van den Bosch, F. *The Study of Plant Disease Epidemics*; American Phytopathological Society: St. Paul, MN, USA, 2007.

16. Alonso Chavez, V.; Parnell, S.; van den Bosch, F. Monitoring Invasive Pathogens in Plant Nurseries for Early-Detection and to Minimise the Probability of Escape. *J. Theor. Biol.* **2015**. submitted for publication.

17. Parnell, S.; Gottwald, T.R.; Gilks, W.R.; van den Bosch, F. Estimating the incidence of an epidemic when it is first discovered and the design of early detection monitoring. *J. Theor. Biol.* **2012**, *305*, 30–36.

18. Parnell, S.; Gottwald, T.R.; Cunniffe, N.J.; Alonso Chavez, V.; van den Bosch, F. Early-detection surveillance for an emerging plant pathogen: A rule of thumb to predict prevalence at first discovery. *Proc. R. Soc. B.* **2015**, *282*, 20151478.

19. Gross, A.; Holdenrieder, O.; Pautasso, M.; Queloz, V.; Sieber, T.N. Hymenoscyphus pseudoalbidus, the causal agent of European ash dieback. *Mol. Plant Pathol.* **2014**, *15*, 5–21.

20. Kirisits, T. Ash dieback caused by Hymenoscyphus pseudoalbidus in a seed plantation of Fraxinus excelsior in Austria. *J. Agric. Ext. Rural Dev.* **2012**, *4*, 184–191.

21. Enderle, R.; Peters, F.; Nakou, A.; Metzler, B. Temporal development of ash dieback symptoms and spatial distribution of collar rots in a

provenance trial of *Fraxinus excelsior*. *Eur. J. For. Res.* **2013**, *132*, 865–876.

22. Bengtsson, S.B.K.; Barklund, P.; von Brömssen, C.; Stenlid, J. Seasonal pattern of lesion development in diseased*Fraxinus excelsior* infected by *Hymenoscyphus pseudoalbidus*. *PLoS ONE* **2014**, *9*, 1–9.

23. Gottwald, T.R.; Schubert, T.S.; Graham, J.H. Citrus Canker : The Pathogen and its impact. *Plant Heal. Prog.* **2002**, *1993*.

Chapter 9

Coordination of Diagnostic Efforts in the Great Plains: Wheat Virus Survey and Modeling of Disease Onset

Mary Burrows[1], Carla Thomas[2], Neil McRoberts, Richard M. Bostock[3], Len Coop, James Stack[4],

[1]Plant Sciences and Plant Pathology Department, Montana State University, Bozeman, MT 59717;

[2]Plant Pathology Department, University of California-Davis, Davis, CA 95616;

[3]Integrated Plant Protection Center, Oregon State University, Corvallis, OR 97331; and

[4]Department of Plant Pathology, Kansas State University, Manhattan, KS, 66506

ABSTRACT

Following the discovery of two new wheat virus diseases in the United States, the Great Plains region (Colorado, Kansas, Montana, Nebraska, North Dakota, Oklahoma, South Dakota, Texas, and Wyoming) of the National Plant Diagnostic Network (NPDN) initiated a project to measure the prevalence of five wheat diseases using indirect ELISA. *Wheat streak mosaic virus* (WSMV), *Wheat mosaic virus* (WMoV), and *Triticum mosaic virus* (TriMV) were found in all nine states. WSMV was the most prevalent, averaging 23 to 47% of samples each year. TriMV and WMoV were detected with WSMV (in up to 76% of the samples). All three mite-transmitted viruses were present in 26% or fewer of the samples. Aphid-transmitted viruses in the barley yellow dwarf complex *Barley yellow dwarf virus*, and *Cereal yellow dwarf virus*-RPV were less frequent (fewer than 65% of the samples). This paper presents the first case-control methodology paper using plant diagnostic laboratory data and the first signed diagnostic data-sharing agreement between the NPDN and its regulatory stakeholders. Samples collected when <700 cumulative degree-days base 0°C, were twice as likely to be virus negative. This proof-of-concept effort highlights the potential of the NPDN and its National Data Repository to develop knowledge about emerging diseases.

Wheat is a staple grain crop for the majority of the world's human population. The United States accounted for 9% of global production in 2012, harvesting 61.8 million metric tons (t) (FAOSTAT 2014), and is often the world's largest wheat exporter, exporting half of its annual yield (Leifert 2013). Viruses are major perennial threats to wheat production in the Great Plains region of the United States. Accurate evaluations of crop loss are difficult to make, but researchers in Kansas estimated losses from wheat viral diseases to be ~1% annually (Appel et al. 2007). In 2011 and 2012, losses were estimated at 4.4 and 3.5%, respectively (Appel et al. 2015). Using a conservative estimate of a 1% yield loss on 20% of the current U.S. production of 60 million t (USDA-NASS 2014) for wheat valued at $0.216/t ($8/bu), U.S. farmers lose $35 million per year due to cereal viruses.

1. WHEAT VIRUSES, VECTORS, AND MANAGEMENT

Increasing global demand for wheat has encouraged growers to employ continuous cropping systems in response to the economic pressures related to the demand for cereals for the use of hay and grain. Continuous cropping has extended the "green bridge," the presence of green plants, including crops, volunteer crops, and grassy weeds between the harvesting of one crop and the planting of the next. Green bridge plants are important reservoirs of cereal viruses (Derksen et al. 2002; Edwards et al. 2011; Hansen et al. 2012). Therefore, cereal viruses are expected to increase in prevalence and remain an economic concern for growers. In addition, the discovery of two new viruses, *Triticum mosaic virus* (TriMV) (Seifers et al. 2008) and *Wheat mosaic virus* (WMoV; formerly High Plains virus [Seifers et al. 1997]), compound the threat of virus diseases in cereals. In 2006, TriMV was identified in varieties of wheat in Kansas resistant to *Wheat streak mosaic virus* (WSMV).

Both WMoV and TriMV cause patterns of chlorosis similar to WSMV. All three diseases are transmitted by the wheat curl mite (WCM; *Aceria tosichella* Keifer [Seifers et al. 2008]). These viruses represent a diverse phylogeny. WSMV is the type member of the genus*Tritimovirus* in the family *Potyviridae* (Stenger et al. 1998). TriMV is in the proposed genus*Susmovirus*, family *Potyviridae* (Fellers et al. 2009). TriMV has 23.2% identity to WSMV, and is most closely related (45.9%) to *Sugarcane streak mosaic virus* – strain AP (Fellers et al. 2009). The taxonomy of WMoV is complex, and it has recently been assigned to genus *Emaravirus* (Skare et al. 2006; Tatineni et al. 2014). WMoV is a multipartite, negative sense, single-stranded RNA virus with eight genomic segments (Tatineni et al. 2014). It has membrane-bound, spherical virions making it more similar to *Bunyaviridae*than *Potyviridae*.

Aphid-transmitted viruses, particularly the barley yellow dwarf (BYD) virus complex are common, cause additional disease losses, and can be easily misdiagnosed based on symptoms (Fig. 1) (Lister and Ranieri 1995). Both BYDV-PAV and CYDV-RPV belong to the family *Luteoviridae*, genus *Luteovirus* (Ali et al. 2014) and are transmitted in a circulative, nonpropagative manner (Gray and Gildow 2003).

Management of mite- and aphid-transmitted viruses is achieved primarily by managing the green bridge and through host plant resistance. In the southern Great Plains, however, reliance on grazed winter wheat makes eliminating the green bridge difficult (Tanaka et al. 2005). Although insecticides exist for aphid management, they are not always economical for use in cereals (Kennedy and Connery 2012). No effective pesticides (acaricides) are labeled for mites in wheat. Cereal crop breeders have worked for years to develop virus-, mite-, and aphid-resistant varieties, but have encountered many challenges. Varietal resistance provides partial protection from WSMV (Graybosch et al. 2009; Lu et al. 2011) and its vector, the WCM (Harvey et al. 1999), but this protection is limited by several factors. Resistant varieties tend to produce lower yields and the WSMV resistance gene currently deployed is not effective above 21°C (Bockus et al. 2001; Lu et al. 2011; Seifers et al. 2007). Additionally, WSMV-resistant varieties are not adapted to northern state climates, so are only grown in Texas, Oklahoma, Kansas, and Nebraska (Byamukama et al. 2011, 2013). Furthermore, virus and mite genomes can change rapidly and overcome host resistance (Harvey et al. 2003; Seifers et al. 2007). Despite these challenges, researchers continue to pursue the development of new, resistant varieties (Rupp et al. 2014).

Figure 1. Symptoms of *Wheat streak mosaic virus* **(A)**, *Triticum mosaic virus* **(B)**, and *Wheat mosaic virus* **(C)** are very similar. *Barley yellow dwarf virus* symptoms are different **(D)**. Virus infections can be confused with nutrient deficiency and other environmental stresses**(E)**.

2. NATIONAL PLANT DIAGNOSTIC NETWORK

Accurate identification of wheat viruses requires diagnostic testing in a laboratory. The National Plant Diagnostic Network (NPDN) was created in 2002 with support from the United States Department of Agriculture National Institute of Food and Agriculture (USDA-NIFA). It is a consortium of diagnostic laboratories at land grant universities and state departments of agriculture that has strengthened partnerships among diagnosticians, university cooperative extension personnel, and regulatory officials (Stack et al. 2006,2014). The NPDN is composed of five regions: the Northeastern Plant Diagnostic Network (NEPDN), the North Central Plant Diagnostic

Network (NCPDN), the Southern Plant Diagnostic Network (SPDN), the Great Plains Plant Diagnostic Network (GPDN), and the Western Plant Diagnostic Network (WPDN).

NPDN is an important part of the U.S. effort to ensure the security of the nation's plant resources (Stack et al. 2006, 2014). Importantly, the NPDN helps inform cooperative extension personnel, researchers, and regulatory officials of emerging threats posed by plant pests and pathogens. The Center for Environmental and Regulatory Information Systems (CERIS) at Purdue University houses and supports the NPDN National Data Repository, which contains the diagnostic results of NPDN laboratories.

The NPDN has created a secure, up-to-date national archive for diagnostic records from participating laboratories throughout the United States and U.S. territories. Authorized specialists can consult the repository to monitor diagnostic trends for common pests and outbreaks of new or unusual pests and diseases in real time.

Diagnosticians across the Great Plains have expressed concern about the accuracy of identifying virus species since the detection of two new mite-transmitted wheat viruses. Symptoms for WMoV and TriMV are virtually indistinguishable from WSMV symptoms. Additionally, BYD symptoms are nearly indistinguishable from abiotic stress or nutrient deficiencies (Bockus et al. 2010). During a meeting in 2007 of the GPDN, diagnosticians discussed the increasing incidence of virus infections in wheat. In response to concerns about accurate diagnosis of virus diseases of cereals and the impact of multiple viral infections on the crops, diagnostic labs of the GPDN began a region-wide collaboration to address the problem (Burrows et al. 2009). The approach was four-pronged: (i) conduct a survey in the Great Plains for the five wheat viruses from 2008 to 2010; (ii) develop and implement a multistate, formal agreement among regulatory agencies to allow diagnosticians to share their diagnostic results with other diagnosticians and regulatory officials; (iii) use the information and analytical tools in the NPDN National Data Repository to understand the timing of detectable wheat virus onset; and iv) use a model based on degree-days as an indicator of optimal timing to conduct sampling and processing during the survey. In this paper, we describe results of this survey and the multistate collaborations enabling new insights into the dynamics of wheat virus pathosystems in the Great Plains.

3. THE GREAT PLAINS WHEAT VIRUS SURVEY AND SAMPLE COLLECTION

This project was a three year collaboration (2008 through 2010) among agricultural extension agents, extension specialists, state departments of agriculture, federal regulators, crop advisors, and NPDN laboratories in nine states: Colorado, Kansas, Montana, Nebraska, North Dakota, Oklahoma, South Dakota, Texas, and Wyoming (Burrows et al. 2009). Samples of winter and spring wheat (*Triticum aestivum* L.) and barley (*Hordeum vulgare* L.) were collected and submitted to the NPDN laboratories. All samples had virus-like symptoms, were part of routine survey activities, and/or were solicited from county extension agents and crop advisors. The samples were not collected according to standardized or random sampling plans, but usually from symptomatic plants. More than 95% of the samples were from winter wheat. Less than 5% of samples were from barley and spring wheat submitted from Montana, North Dakota, and South Dakota. Sample collection was not randomized and collection criteria were not consistent from state to state. Samples were collected and submitted to NPDN diagnostic laboratories based on perceived grower need and/or surveys conducted by individual cooperating states.

4. VIRUS DETECTION

All wheat samples received in GPDN diagnostic laboratories were tested for five viruses: WSMV, TRIMV, WMoV, *Barley yellow dwarf virus*-PAV (BYDV-PAV), and *Cereal yellow dwarf virus*-RPV (CYDV-RPV) using indirect ELISA (Agdia Diagnostics, Elkhart, IN). An agreement among diagnostic laboratories was created to standardize wheat virus testing throughout the GPDN laboratories. Samples were individually processed according to the instructions of the manufacturer of the ELISA kits. Each kit consisted of a single noncoated ELISA plate (96-well formatted in 8-well break-apart 'strips') for each of the five viruses, appropriate positive and negative controls, and reagents required for the assays including antibodies and buffers. Absorbance at 405 nm was measured using an ELISA plate reader, which varied by manufacturer and model among diagnostic laboratories. Samples were considered positive if the absorbance value of the sample exceeded twice the absorbance value of the negative control (i.e., healthy wheat). If the ELISA was inconclusive the sample was considered negative. TriMV test results from North Dakota and South Dakota in 2009 were excluded from the data set due to difficulties with the TriMV ELISA. However, end-point PCR and BLAST analysis confirmed the presence of TriMV in both states. At the time, PCR tests were nonstandardized, using a variety of primers derived from the currently known sequences from GenBank. Infected plant tissues were preserved by lyophilization and storage at –80°C for use in future research projects (Rogers 2011).

5. SURVEY DATA REPORTING

A total of 2,456 samples were collected and processed during the course of the survey (Table 1). Data were expressed as the percentage of samples infected compared with the number of samples processed for each county for each year. The survey sample set reported in this paper is biased due to the nature of sample collection. Therefore, it is difficult to draw conclusions regarding year-to-year variation in virus frequency since sample numbers were largely tied to survey efforts in individual states. For example, in 2009 in Montana, a cooperative agricultural pest survey of small grains was performed with the Montana Department of Agriculture and APHIS-PPQ. This resulted in more small grains samples submitted for virus diagnosis than would normally come into the diagnostic laboratory.

Despite these limitations, this dataset provided confirmation that WSMV was the most prevalent, meaning the most commonly occurring virus as quantified by incidence, detected in every state. Overall, 36% of sample date and location combinations tested positive for WSMV with an incidence ranging from 6 to 83% among states (Table 1). TriMV was the next most prevalent virus detected with incidence averaging 16% among states and ranging from 0 to 57%. Thirteen percent of samples tested positive for WMoV (range 0 to 41%) and 13% tested positive for BYDV-PAV (range 0 to 65%). CYDV-RPV was the least prevalent virus detected (3% overall average, range 0 to 14%). Forty-six percent of samples had no detectable viruses and 30% of samples tested positive for only one virus. Twenty-one percent of samples were coinfected, meaning they tested positive for two or more viruses (Tables 1 and 2).

Both WMoV and TriMV were found in all nine states during the course of the survey. TriMV had not been detected outside of Kansas prior to this survey. As specified by the NPDN Chain of Communication-Chain of Custody Standard Operating Procedure, State Plant Regulatory Officials (SPROs) and State Plant Health Directors (SPHDs) were notified in each state by the

NPDN diagnostician, upon the first positive results of these new viruses. From the widespread nature of this newly identified virus, it appears to have been present but unrecognized in U.S. cropping systems for an extended period of time. TriMV was therefore not designated as a "regulated" pest by APHIS-PPQ. If a pest is designated as "regulated," it is not generally established in the United States, is frequently intercepted from commodities at U.S. ports of entry, and has the potential to cause serious economic or environmental damage in the United States. (USDA-APHIS 2012). Lack of recognition of TriMV as a distinct virus may have been due to the similarity of TriMV symptoms to WSMV. The current survey results also confirmed that WSMV was more common than TriMV and that TriMV was found primarily in coinfections with WSMV, rather than as a single infection (Byamukama et al. 2013) (Table 1). Since the inception of this survey, others have also identified TriMV in samples outside of Kansas and its occurrence appears to be widespread (Byamukama et al. 2013).

We also observed that TriMV and WMoV were highly prevalent (7 to 20% of samples on average per year), and often present in coinfections with other WCM-transmitted viruses (Table 1). All three WCM-transmitted viruses were rarely found in a single plant sample (an average of 1 to 5% of samples per year; Table 2).

Aphid-transmitted viruses in the BYD complex (BYDV-PAV and CYDV-RPV) were infrequent. Possible reasons for this include: (i) a less frequent occurrence in the Great Plains than mite-transmitted viruses, (ii) growers and agricultural professionals were more confident in their ability to recognize this virus disease and therefore did not submit samples for diagnosis, or (iii) symptoms of BYD complex diseases are easily mistaken for nutrient deficiencies. During the years of the study BYD was a minor disease of wheat. In 2014 there were outbreaks of BYD in Idaho, North Dakota, and Minnesota (Friskop 2014;Marshall and Rashed 2014).

Table 1. Incidence (%) of wheat samples testing positive for *Wheat streak mosaic virus*, (WSMV), *Wheat mosaic virus* (WMoV),*Triticum mosaic virus* (TriMV), *Barley yellow dwarf virus*(BYDV)-PAV, and *Cereal yellow dwarf virus* (CYDV)-RPV, and coinfections during a survey of Great Plains states, 2008 to 2010

		Percent of samples testing positive using ELISA[a,b]									
		Mite transmitted			Aphid transmitted		Coinfections				
State	n	WSMV	WMoV	TriMV	BYDV-PAV	CYDV-RPV	WSMV + WMoV	WSMV + TriMV	WMoV + TriMV	All mites[c]	All aphids[d]
2008											
Colorado	51	61	10	10	10	4	8	8	0	0	0
Kansas	53	62	38	30	6	2	15	21	13	8	0
Montana	23	43	9	0	9	0	9	0	0	0	0
Nebraska	66	39	8	27	5	3	8	18	5	5	2
North Dakota	44	40	12	0[b]	2	0	9	0	0	0	0
Oklahoma	93	27	30	6[b]	16[b]	3 [b]	16	4	3	3	2
South Dakota	96	28	7	2	3	0	7	2	1	1	0
Texas	307	83	41	57	14	2[b]	37	53	28	26	0
Wyoming	21	38	19	24	0	0	5	10	10	0	0
Average		47	19	17	7	2	13	13	13	5	0
Total	754										
2009											
Colorado	145	50[b]	24[b]	21[b]	32[b]	8[b]	70	67	46	6	5
Kansas	50	14	0	2	10	0	14	16	2	0	0
Montana	149	29	4[b]	7	1	2	3	5	1	1	1
Nebraska	97	14	16	44	8	14	31	57	60	3	1
North Dakota	95	52	12	.	6	9[b]	63	.	.	.	0
Oklahoma	77	49	17	18[b]	65	4	66	68	36	0	4
South Dakota	44	27	14	.	5	.	41	.	.	.	.
Texas	320	44	4	14	20	0	48	58	18	2	0
Wyoming	21	10	19	0	5	0	76	14	10	0	0
Average		32	12	20	17	4	46	46	29	2	1
Total	998										
2010											

Colorado	118	44	14	25	23	11	5	20	2	2	3
Kansas	62	13	0	8	18	0	0	5	0	0	0
Montana	61	48	4[b]	0[b]	0[b]	0[b]	2	0	0	0	0
Nebraska	122	11	6	47[b]	9	0	0	0	0	0	0
North Dakota	35	60	15	0	8	0	15	0	0	0	0
Oklahoma	36	33	22	11	50	8	17	11	11	11	8
South Dakota	67	13	4	4	9	1	1	0	0	0	1
Texas	170	26	1	8	13	6	1	8	1	1	3
Wyoming	33	6	0	0	0	0	0	0	0	0	0
Average		28	7	11	14	3	4	5	1	1	2
Total	704										

[a] A period (.) indicates difficulties with the ELISA for detecting virus; data not shown.

[b] Percentage based on total number of samples per state (n) unless noted here. In 2008, North Dakota ($n = 44$) processed 41 TriMV samples; Oklahoma ($n = 93$) processed 89 TriMV, 90 BYDV-PAV and 90 CYDV-RPV samples; and Texas ($n = 307$) processed 257 BYDV-RPV samples. In 2009, Colorado ($n = 145$) processed 144 WSMV, 100 WMoV, 98 TriMV, 92 BYDV-PAV, and 91 CYDV-RPV samples; Montana ($n = 149$) processed 148 WMoV samples; North Dakota ($n = 95$) processed 93 CYDV-RPV samples; Oklahoma ($n = 77$) processed 76 TriMV samples. In 2010, Montana ($n = 61$) processed 46 samples for WMoV and 47 samples for TriMV, BYDV-PAV, and CYDV-RPV; Nebraska ($n = 35$) processed 15 samples for TriMV.

[c] Percentage of plants tested positive with ELISA for mite-transmitted viruses (WSMV, WMoV, and TriMV).

[d] Percentage of plants tested with ELISA that were positive for aphid-transmitted viruses (BYDV-PAV and CYDV-RPV).

6. DEVELOPMENT OF A DEGREE-DAY MODEL FOR PREDICTING ONSET OF DETECTIBLE WSMV

WSMV records and data analysis tools from the NPDN National Data Repository were used to develop a degree-day model for estimating the onset of detectable WSMV in field samples. Degree-day map layers were imported into the NPDN National Data Repository analysis system from the Oregon State University Integrated Plant Protection Center's (OSU IPPC) weather-based online decision support platform via the website http://uspest.org/wea (Oregon State University 2012). For each sample, cumulative degree-days base 0°C (CDD0°C) were calculated from January 1 each year to the date of sample collection. CDD0°C was selected to be consistent with wheat phenology (growth stage) models postulating that phenology may be correlated with detectable WSMV. Wheat phenology models were selected that are well documented and currently in use (Hay and Kirby 1991; Karow et al. 1993; Keppler et al. 1988; McMaster and Wilhelm 1997; McMaster and Smika 1988; Miller et al. 2001; Porter and Gawith 1999; Saiyed et al. 2009; Streck et al. 2003; Wang and Engel 1998) Degree-day values were calculated using the temperature-based single sine curve method with no upper threshold (Baskerville and Emin 1969). The start date of January 1 was selected because McMaster and Smirka determined that winter wheat phenology degree-day models that started at January 1 more accurately estimated the correct growth stage than models that started at sowing or emergence for winter wheat since vernalization and other synchronization mechanisms in winter wheat reduced the variability contributed by fall planting dates particularly in northern latitudes (McMaster and Smika 1988; Streck et al. 2003; Wang and Engel 1998). Since sample location was collected at the county resolution but not subcounty, the geographic center of the county (county centroid) was used as the sample location. The OSU IPPC's weather-based decision support platform was used to generate a CDD0°C value for the nearest weather station to the county centroid on the date of collection for each sample. The system's weather database of over 16,000 weather stations from numerous public networks was obtained from Mesowest at the University of Utah and from numerous grower-led weather networks (University of Utah 2012). After ingestion of the raw weather data, the decision support platform performed a rigorous automated quality control system that scans for and corrects errors and missing values using distance and elevation weighted regression estimation of values from nearby weather stations (unpublished data).

Degree-day values for the sample collection dates in 2008, 2009, and 2010 were plotted for each sample to estimate a threshold value for onset of detectible WSMV (Fig. 2). Only 30 of the 435 samples which tested positive for WSMV were collected below 700 CDD0°C. It is possible that prior to reaching the 700 CDD0°C threshold there is a phenological development in the host plant or vector that prevents virus titer from multiplying to detectable levels in the host. Assuming a range of 600 to 800 CDD0°C between fall planting and January 1, the growth stage of winter wheat at 700 CDD0°C after January 1 is between the "6-leaves-expanded-with-tillering-ongoing" growth stage (Feekes wheat model growth stage 6 or Zadocks wheat model growth stage 31) and "flag-leaf-sheath-extended" growth stage (Feekes wheat model growth stage 10 or Zadocks wheat model growth stage 45) (Karow et al. 1993; Keppler et al. 1988; Large, 1954; Miller et al. 2001;Zadocks et al. 1974). This is when the plant is in late vegetative stages and shifting its metabolic processes toward reproductive stages (heading). The development and growth rate of wheat organs (e.g., leaves, tillers, spikelets, and inflorescences) changes as do minimum, optimum, and maximum base temperatures when the wheat plant transitions from the vegetative to reproductive stages (Porter and Gawith 1999; Slafer and Rawson 1995; Streck et al. 2003). It is not known if the mechanism of symptom development or virus titer is affected by this shift.

A regional diagnostic data-sharing agreement was developed among GPDN diagnosticians, state regulatory officials, SPHDs, and SPROs that allowed GPDN diagnosticians to share diagnostic records. All SPROs from GPDN states signed this agreement allowing GPDN diagnosticians to share the diagnostic results with all GPDN SPROs, SPHDs, and diagnosticians. This is the first time a regional diagnostic data-sharing agreement was developed among NPDN diagnosticians and state and federal regulatory officials to allow this degree of sharing. Prior to this, diagnosticians were not allowed to share diagnostic results with each other between states. In 2008 and 2009, data were shared on weekly teleconference calls with diagnosticians.

During 2009, a degree-day threshold of 700 CDD0°C was proposed to indicate the first detection of WSMV positive samples each year. The threshold was based on the results from the 2008 and 2009 diagnostic samples. In 2010, county resolution maps of observed WSMV-positive and WSMV-not-detected counties and the continuous variable CDD0°C values were generated using the NPDN analysis tools in the NPDN National Data Repository (Fig. 3). A county was considered WSMV positive if any sample results from the county were WSMV positive to date since January 1 of the same year. Similar county resolution maps were generated for the other four wheat virus diseases surveyed, but were not integrated into a degree-day map. These multistate maps were distributed by e-mail to the diagnosticians, SPROs, and SPHDs as semimonthly updates during April, May, June, and July 2010 along with a text summary of diagnostic results over the prior two weeks (Supplementary Fig. S1).

Table 2. Incidence (%) of single and coinfection of wheat plants during a survey of Great Plains states, 2008 to 2010

		Number of viruses coinfecting each sample					
		0	1	2	3	4	5
2008	n^a	Percent of samples infected (%)[b]					
Colorado	51	22	63	16	0	0	0
Kansas	53	8	55	30	8	0	0
Montana	23	48	43	9	0	0	0
Nebraska	66	47	29	20	5	0	0
North Dakota	44	57	34	9	0	0	0
Oklahoma	93	49	28	17	3	2	0
South Dakota	96	68	23	7	1	0	0
Texas	307	8	23	37	28	4	0
Wyoming	21	43	33	24	0	0	0
Average		39	37	19	5	1	0
		Number of viruses coinfecting each sample					
		0	1	2	3	4	5
2009	n	Percent of samples infected (%)[b]					
Colorado	145	33	41	14	11	0	1
Kansas	50	82	15	3	0	0	0
Montana	149	67	24	7	1	0	0
Nebraska	97	36	41	17	6	0	0
North Dakota	95	39	48	8	4	0	0
Oklahoma	77	10	34	48	8	0	0
South Dakota	44	66	23	11	0	0	0
Texas	320	46	34	14	4	2	0
Wyoming	21	76	14	10	0	0	0
Average		49	29	18	4	0	0
		Number of viruses coinfecting each sample					
		0	1	2	3	4	5

2010	*n*	Percent of samples infected (%)[b]					
Colorado	118	26	38	29	6	1	0
Kansas	62	69	24	5	2	0	0
Montana	61	52	48	2	0	0	0
Nebraska	35	57	40	3	0	0	0
North Dakota	122	34	49	16	1	0	0
Oklahoma	36	17	53	19	11	0	0
South Dakota	67	72	25	1	1	0	0
Texas	170	61	26	10	2	1	0
Wyoming	33	94	6	0	0	0	0
Average		54	34	9	3	0	0

[a] Number of cereal plant samples tested from each state for *Wheat streak mosaic virus*, *Wheat mosaic virus*, *Triticum mosaic virus*, *Barley yellow dwarf virus*-PAV, and *Cereal yellow dwarf virus*-RPV.

[b] Data expressed as percent of samples from each state tested by ELISA that were infected with 0 to 5 viruses in a single sample.

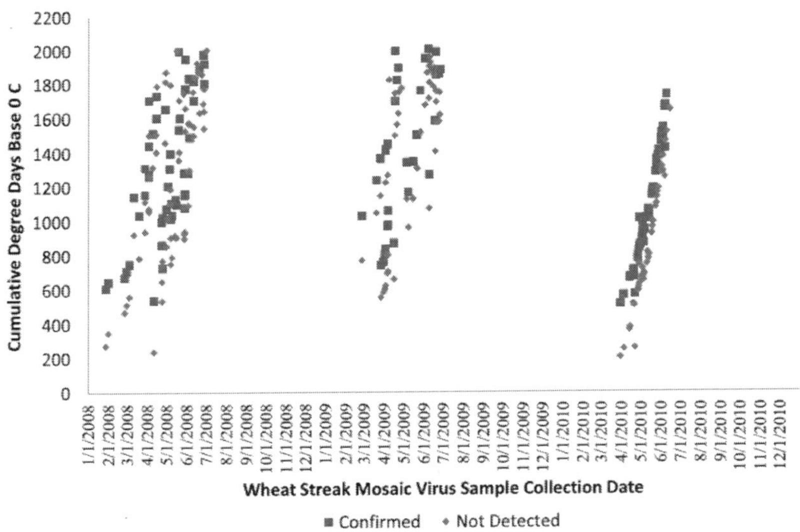

Figure 2. Plot of cumulative degree-days base 0°C (CDD0°C) calculated from January 1 until the wheat sample date compared with *Wheat streak mosaic virus* (WSMV) diagnostic test result (confirmed positive or not-detected) for sample collection dates in 2008, 2009, and 2010 for winter wheat in the Great Plains states, United States, in from 2008 to 2010. Only 30 of the 435 samples which tested positive for WSMV were collected above 700 CDD0°C.

7. NPDN AS A TOOL FOR DISSEMINATING INFORMATION: 2010 WHEAT VIRUS ALERTS

The 700 CCD0°C model was able to give general guidance on when to expect the first positive WSMV samples to originate from a local area (Fig. 3). The first positive WSMV county detections (red outlined counties on May 28, 2010) were observed within 2 weeks after CDD0°C values were 700 to 900 (the pale blue locations on the map indicated by red arrows and circles on May 14th insert). When CDD0°C was 700 to 900 and onset of disease was not observed, it may have been due to an absence of the vector or the virus. Based on these observations, a more rigorous statistical analysis was performed and is presented in the case-control section below.

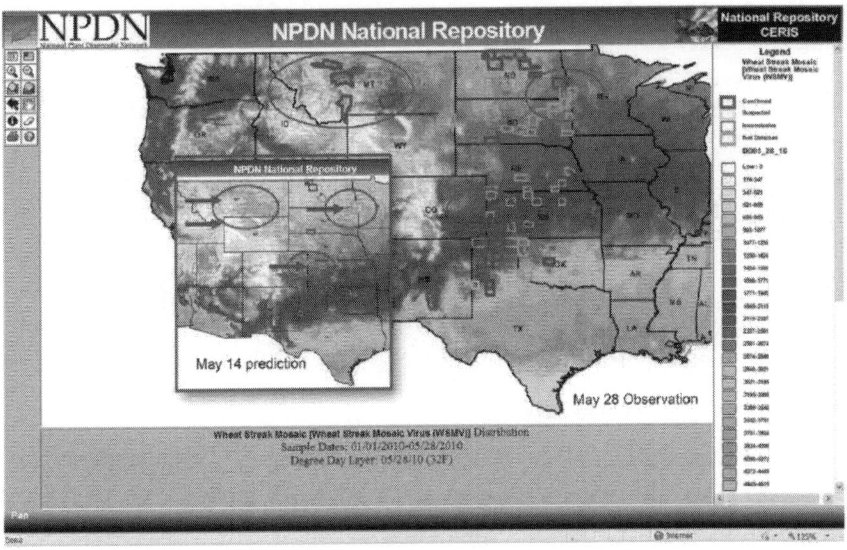

Figure 3. Example of the relationship over time between cumulative degree-days base 0°C (CDD0°C) since January 1 and the first county positive *Wheat streak mosaic virus* (WSMV) test result of the year. Arrows indicate areas between 700 and 900 CDD0°C since January 1, on May 14, 2010 predicting onset of WSMV symptoms by May 28, 2010. Red circles indicate first observed WSMV positive test results on May 28, 2010 by county in predicted areas.

Sample collection patterns for WSMV in 2010 were different than in 2008 and 2009. Two discrete populations were observed in 2010 when CDD0°C was plotted by date and diagnostic test result for samples collected prior to 2,200 CDD0°C (Fig. 2). For any CDD0°C value in 2010, positive samples were detected before not-detected samples for all CDD0°C values with positive WSMV detections. In addition, 2010 data had a smaller range of sample dates than the previous years. It is possible that the wheat virus updates influenced the timing of sample collection and processing, shortening the sampling period to be closer to the time of the observed first occurrence of WSMV positive test results.

8. CASE-CONTROL ANALYSIS

In 2012, a case-control analysis was conducted to test the relative value of using the 700 CDD0°C threshold compared with other CDD0°C thresholds. The NPDN National Data Repository held diagnostic results for samples collected from the GPDN from 2006 to 2011 at the time that the case-control analysis was conducted in 2012. In the case-control study, replicated samples were pooled as one sample for each location, date, and test result combination to reduce the bias when many samples were collected at one sample point, since only one sample was collected from many locations. In the case-control analysis, there were 849 unique degree-day-location-sample-date combinations for WSMV, 414 (49%) samples tested as WSMV not-detected and 435 (51%) samples tested as WSMV positive.

The association between degree-day accumulation and virus incidence was examined with the case-control design method (Hughes et al. 1999) and discussed at length by others (Hughes and McRoberts 2014; Madden et al. 2007; Yuen and Hughes 2002). In this study, our objective was to test the potential for using CDD0°C as an indicator of final ELISA outcome for WSMV for an individual plant sample. In other words, could degree-day accumulation provide a simple indicator of ELISA outcome that can be used as a prior proxy for, or in addition to, diagnostic laboratory tests? If so, then the degree-day value measured by CDD0°C could inform sample collectors as to the most likely first date to detect virus-positive samples, improving survey efficiency.

To examine the accuracy of degree-day based predictions, it is necessary to have an independent, "gold standard" classification of samples to compare with predicted virus status. The WSMV ELISA test result was used as the "gold standard" for the presence of WSMV in the plant sample.

To determine the CDD0°C threshold with the highest probability of obtaining a virus-positive ELISA test from a field sample, the CDD0°C for each sample was partitioned at different threshold values in 100 CDD0°C intervals starting at 500 CDD0°C (Hughes et al. 1999). Each sample was classified as either having accumulated fewer than (<) or greater than or equal to (≥) the threshold CDD0°C at values of 500, 600, 700, 800, 900, and 1,000 CDD0°C. For each cumulative degree-day threshold, a 2 × 2 outcome table was constructed in which every sample was cross-classified by whether it was a "case" (V+) or "control" (V–).

A threshold of 700 CDD0°C provided the largest amount of formal information concerning the virus status of test samples added by a sample's CDD0°C value (Table 3). This result retrospectively concurs the use of the 700 CDD0°C threshold selected from the 2008–2009 diagnostic results and distributed during the 2010 wheat virus alerts.

Table 3. Contingency table (number of samples) and resulting likelihood ratios in a case-control analysis for a threshold of 700 cumulative growing degree-days base 0°C (CDD0°C) starting January 1 for *Wheat streak mosaic virus* (WSMV) ELISA test results during 2008 to 2010 in the Great Plains, United States. Each sample represents a unique county-date-year combination

	Observed no. samples			
ELISA results	CDD0°C <700	CDD0°C >700	Total no. samples	Likelihood ratio
WSMV not detected	56	357	413	Case LR+ 1.077 Case LR- 0.509
WSMV positive	30	405	405	Control LR- 0.928 Control LR+ 1.966
Totals	85	762	848	

Samples collected when CDD0°C were less than 700 were twice as likely to be virus negative and half as likely to be virus positive. This result supports the hypothesis that it is most productive to sample when CDD0°C are at or above 700 if the goal is to detect positive results. While these may appear to be relatively modest gains in accuracy at the level of an individual sample, they may represent significant gains in efficiency obtained by better timing of sample collection when considered at the scale of all samples collected over the entire region over multiple years.

Case-study methodology was originally applied to clinical diagnostic testing of individual patients in medical clinics. Its use in plant health was adapted to plant populations in the field, not to individual plants. This is the first application of a case-control design to plant diagnostic clinic results for individual plants and uses a multistate plant-diagnostic data set.

Even after taking into account degree-day accumulation, the results indicate the possible effect of other factors such as plant variety, planting date, and prior field history. These factors were not analyzed and could also influence the odds that the sample test result will be WSMV positive or WSMV not-detected. There was no standardization in sampling strategy with regards to host variety, host susceptibility, prior WSMV or WCM presence for the collection site or neighboring collection sites, water availability, plant fertility, weed control, or presence of a green bridge. These factors are known to impact phenology stage and also may be important in determining ELISA results at or above 700 CDD0°C (McMaster and Smika 1988; Slafer and Rawson 1995). Degree-day calculations started on January 1 to minimize the additional impacts of planting date, winter minimum temperatures, winter snow cover due to vernalization, and other synchronization factors. The degree-day model is likely an estimate of the stage of wheat phenology when symptoms are first expressed (Karow et al. 1993; Keppler et al. 1988; Miller et al. 2001;Porter and Gawith 1999; Streck et al. 2003; Wang and Engel 1998). Future work should explore likelihood ratios after dividing the data set into geographic subregions, with similar planting dates, cultivars, and other regional factors. These may have a practical effect on the likelihood ratios for WSMV in individual plants, particularly for virus-positive outcomes.

9. PERSPECTIVES SINCE THE SURVEY

This project was a prime example of taking a common concern, engaging a network of collaborators, and tapping diagnostic resources to rapidly learn important information about an emerging pathosystem. Results were used to leverage over $5 million in funding for additional research, extension, and educational efforts that will directly impact the profitability of farms in the Great Plains. Despite the known inconsistent methods of data collection, the initial survey provided a useful starting point for a larger effort. Since this survey, research scientists and extension personnel have cooperated on a number of wheat virus projects to understand the pathosystem and mitigate disease risk for growers. This cooperation has culminated in a six-state project supported by USDA-NIFA to develop a risk model for disease throughout the Great Plains, from Texas to Montana. Core experiments being conducted in Montana, Nebraska, and Texas include a variety trial, planting date study, and monitoring of important green bridge hosts for the presence of virus and potential mite vectors. These data along with weather data and the further ongoing work on the contribution of hail to subsequent disease, will feed into a risk model that growers can use (Byamukama et al. 2011, 2013).

The wheat virus system also is being used in educational modules for K-12 education and outreach. We developed a series of interactive approaches that have been used at outreach presentations including the use of "gigaballs" as virus models for young children. We have also conducted grower surveys to identify barriers to the adoption of disease management recommendations, such as avoiding green bridges, and to evaluate changing perceptions and grower practices. Some of the most important barriers to eliminating the green bridge are social and marketing pressures, not an understanding of the science. The results of these ongoing projects underscore the importance of science-based extension education of growers and the need for continuing education of new growers, especially as farming operations become larger and more complex. The diagnostic network results were a key component of that educational system.

10. CONCLUSION

This paper summarizes an initial effort to learn more about an emerging complex of wheat viruses in the Great Plains using the resources of the NPDN. Our work constitutes the first signed formal diagnostic data-sharing agreement between NPDN and its regulatory stakeholders. This is also the first paper to apply case-control methodology to clinical plant pathology using plant-diagnostic laboratory data. This proof-of-concept effort highlights the potential of the NPDN and the NPDN National Data Repository to develop useful information about an emerging disease. We learned that WSMV, WMoV, and TriMV were found in all nine Great Plains states and WSMV was the most prevalent. Additionally, TriMV and WMoV were usually found in coinfections with WSMV, but seldom was a plant infected with all three. Aphid-transmitted viruses in the BYD complex

(BYDV-PAV and CYDV-RPV) were relatively infrequent. In 2010, semimonthly maps and tables indicating the presence of these wheat virus diseases demonstrated a potential for influencing efficiency by timing sample collection.

Samples collected when CDD0°C were less than 700 were twice as likely to be virus negative and half as likely to be virus positive. Sampling and diagnostics are resource-intensive processes requiring significant investments of human capital, supplies, and reagents. The gain in likelihood of a positive WSMV test result based on time of sampling has potential to improve sampling and testing efficiency and reduce these costs. The information in the present study would have been very difficult to obtain without the resources of the NPDN. The potential for the resources of the NPDN and its National Data Repository to advance scientific discovery in the context of new disease and pest outbreaks is promising.

ACKNOWLEDGMENTS

The authors wish to acknowledge the Great Plains Diagnostic Network for funding this survey as well as the diagnosticians, plant pathologists, and other personnel who collected and processed plant samples. We would especially like to thank Tamla Blunt, Gary Franc, Dai Ito, Kasia Kinzer, Jennifer Dominiak-Olson, Judy O'Mara, Jacob Price, Charlie Rush, Connie Tande, and Amy Timmerman. This work was funded with a grant from a subcontract under the USDA-CSREES IPM Grant number 2007-37620-18152, USDA-NIFA Grant number 2012-37620-19627, PubMedUSDA-NIFA Grant number 2012-37620-19626.

REFERENCES

1. Ali, M., Hameed, S., and Tahir, M. 2014. Luteovirus: insights into pathogenicity. Arch. Virol. 159:2853-2860.

2. Appel, J., DeWolf, E., Bockus, W., and Bowden, R. L. 2007. Preliminary 2007 Kansas wheat disease loss estimates. Kansas Cooperative Plant Disease Survey Report. Kansas Department of Agriculture, Topeka, KS.

3. Appel, J., DeWolf, E., Todd, T., and Bockus, W. 2015. Preliminary 2015 Kansas wheat disease loss estimates. Kansas Cooperative Plant Disease Survey Report.

4. Baskerville, G. L., and Emin, P. 1969. Rapid estimation of heat accumulation from maximum and minimum temperatures. Ecology 50:514-517.

5. Bockus, W. W., Appel, J. A., Bowden, R. L., Fritz, A. K., Gill, B. S., Martin, T. J.,Sears, R. G., Seifers, D. L., Brown-Guedira, G. L., and Eversmeyer, M. G. 2001.Success stories: Breeding for wheat disease resistance in Kansas. Plant Dis. 85:453-461.

6. Bockus, W. W., Bowden, R. L., Hunger, R. M., Morrill, W. L., Murray, T. D., andSmiley, R. W., eds. 2010. Compendium of Wheat Diseases and Pests, 3rd Ed. APS Press, St. Paul, MN.

7. Burrows, M., Rush, C., and Franc, G. 2009. Occurrence of viruses in wheat in the Great Plains region, 2008. Plant Health Prog. Online: 10.1094/PHP-2009-0706-01-RS.

8. Byamukama, E., Seifers, D. L., Hein, G. L., De Wolf, E., Tisserat, N. A.,Langham, M. A. C., Osborne, L. E., Timmerman, A., and Wegulo, S. N. 2013.Occurrence and distribution of Triticum mosaic virus in the central Great Plains. Plant Dis. 97:21-29.

9. Byamukama, E. Z., Tatineni, S., Hein, G. L., and Wegulo, S. N. 2011. Co-inoculation of wheat with Triticum mosaic virus and Wheat streak mosaic virus exacerbates loss of fresh and dry matter. Phytopathology

10. Derksen, D. A., Anderson, R. E., Blackshaw, R. E., and Maxwell, B. 2002. Weed dynamics and management strategies for cropping systems in the Northern Great Plains. Agron. J. 94:174-185.

11. Edwards, J. T., Carver, B. F., Horn, G. W., and Payton, M. E. 2011. Impact of dual-purpose management on wheat grain yield. Crop Sci. 51:2181-2185.

12. FAOSTAT. 2014. Food and Agriculture Organization of the United Nations database. Online: http://faostat.fao. rg/site/567/Desktop Default .aspx?PageID=567#ancor

13. Fellers, J. P., Seifers, D., Ryba-White, M., and Martin, T. J. 2009. The complete genome sequence of Triticum mosaic virus, a new wheat-infecting virus of the High Plains. Arch. Virol. 154:1511-1515.

14. Friskop, A. 2014. North Dakota State University Integrated Pest Management survey of wheat. Available online https://www. ag. ndsu.edu/ndipm.

15. Gray, S. M., and Gildow, F. E. 2003. Luteovirus-aphid interactions. Ann. Rev. Phytopathol. 41:539-566

16. Graybosch, R. A., Peterson, C. J., Baenziger, P. S., Baltensperger, D. D., Nelson, L. A., Jin, Y., Kolmer, J., Seabourn, B., French, R., Hein, G., Martin, T. J.,Beecher, B., Schwarzacher, T., and Heslop-Harrison, P. 2009. Registration of 'Mace' hard red winter wheat. J. Plant Regist. 3:51-56.

17. Hansen, N. C., Allen, B. L., Baumhardt, R. L., and Lyon, D. J. 2012. Research achievements and adoption of no-till, dryland cropping in the semi-arid US Great Plains. Field Crops Res. 132:196-203.

18. Harvey, T. L., Martin, T. J., and Seifers, D. L. 2003. Resistance to the wheat curl mite (Acari: Eriophyidae) prevents loss in wheat yield. J. Ag. Urban Ent. 20:7-10. [ISI]

19. Harvey, T. L., Seifers, D. L., Martin, T. J., Brown-Guedira, G., and Gill, B. S.1999. Survival of wheat curl mites on different sources of resistance in wheat. Crop Sci. 39:1887-1889.

20. Hay, R. K. M., and Kirby, E. J. M. 1991. Convergence and synchrony-a review of the coordination of development in wheat. Aust. J. Agric. Res. 42:661-700.

21. Hughes, G., and McRoberts, N. 2014. The structure of diagnostic information. Australas. Plant Pathol. 43:267-286.

22. Hughes, G., McRoberts, N., and Burnett, J. 1999. Decision making and diagnosis in disease management. Plant Pathol. 48:147-153.

23. Karow, R. S., Klepper, E. L., Rickman, R. W., Toll, T. R. 1993. Early growth and development of cereals, Oregon State University Extension Service Bull. EM8542.

24. Kennedy, T. F., and Connery, J. 2012. Control of barley yellow dwarf virus in minimum-till and conventional-till autumn-sown cereals by insecticide seed and foliar spray treatments. J. Agric. Sci. 150:249-262.

25. Klepper, B., Rickman, R. W., Zuzel, J. F., and Waldman, S. E. 1988. Use of growing degree days to porject sample dates for cereal crops. Agron. J. 80:850-852.

26. Large, E. C. 1954. Growth stages in cereals. Plant Pathol. 3:128-129.

27. Leifert, O. 2013. U.S. wheat trade. USDA Economic Research Service. Online:http://www.ers.usda.gov/topics/crops/wheat/trade.aspx#.U9q2L_17FxQ

28. Lister, R. M., and Ranieri, R. 1995. Distribution and Economic Importance of Barley Yellow Dwarf. APS Press, St. Paul, MN.

29. Lu, H., Kottke, R., Devkota, R., St.Amand, P., Bernardo, A., Bai, G., Byrne, P.,Martin, J., Haley, S., and Rudd, J. 2011. Molecular mapping of Wsm2 for Wheat streak mosaic virus resistance in winter wheat line CO960293-2. Poster 305, Plant and Animal Genomes XIX Conference, San Diego, CA, Jan 15-19, 2011.

30. Madden, L. V., Hughes, G., and Bosch, F. V. D. 2007. The Study of Plant Disease Epidemics. APS Press, St. Paul, MN.

31. Marshall, J. M. and Rashed, A. 2014. Barley yellow dwarf virus in Idaho cereal crops. University of Idaho Extension Publication CIS1210.

32. McMaster, G. S., and Smika, D. E. 1988. Estimation and evaluation of winter wheat phenology in the central Great Plains. Agric. For. Meteorol. 43:1-18.

33. McMaster, G. S., and Wilhelm, W. W. 1997. Growing degree-days: one equation, two interpretations. Agric. For. Meteorol. 87:291-300.

34. Miller, P., Lanier, W., Brandt, S. 2001. Using growing degree days to predict plant stages, Montana State University Extension Service MontGuide MT200103AG7.

35. Oregon State University. 2012. IPM pest and plant disease models and forecasting. Oregon State University Integrated Plant Protection Center. Online:http://www.uspest.org/wea

36. Porter, J. R., and Gawith, M. 1999. Temperatures and the growth and development of wheat: a review. Eur. J. Agron.

37. Rogers, S. 2011. A decision tool and molecular typing technique for a plant pathogen forensic application, using Wheat streak mosaic virus as a model pathogen. Oklahoma State University, Ph.D. Dissertation.

38. Rupp, J. L. S., Simon, Z. G., Gillett-Walker, B., and Fellers, J. P. 2014. Resistance to Wheat streak mosaic virus identified in synthetic wheat lines. Euphytica 198:223-229.

39. Saiyed, I. M., Bullock, P. R., Sapirstein, H. D., Finlay, G. J., and Jarvis, C. K.2009. Thermal time models for estimating wheat phenological development and weather-based relationships to wheat quality. Can. J. Plant Sci. 89:429-439.

40. Seifers, D. L., Harvey, T. L., Martin, T. J., and Jensen, S. G. 1997. Identification of the wheat curl mite as the vector of the High Plains virus of corn and wheat. Plant Dis. 81:1161-1166.

41. Seifers, D. L., Martin, T. J., Harvey, T. L., Fellers, J. P., Stack, J. P., Ryba-White, M., Haber, S., Krokhin, O., Spicer, V., Lovat, N., Yamchuk, A., andStanding, K. G. 2008. Triticum mosaic virus: A new virus isolated from wheat in Kansas. Plant Dis. 92:808-817.

42. Seifers, D. L., Martin, T. J., Harvey, T. L., and Haber, S. 2007. Temperature-sensitive Wheat streak mosaic virus resistance identified in KS03HW12 wheat. Plant Dis. 91:1029-1033.

43. Skare, J. M., Wijkamp, I., Denham, I., Rezende, J. a. M., Kitajima, E. W., Park, J. W., Desvoyes, B., Rush, C. M., Michels, G., Scholthof, K. B. G., and Scholthof, H. B. 2006. A new eriophyid mite-borne membrane-enveloped virus-like complex isolated from plants. Virology 347:343-353.

44. Slafer, G. A., and Rawson, H. M. 1995. Rates and cardinal temperatures for processes of development in wheat: effects of temperature and thermal amplitude. Aust. J. Plant Physiol. 22:913-926.

45. Stack, J., Cardwell, K., Hammerschmidt, R., Byrne, J., Loria, R., Snover-Clift, K., Baldwin, W., Wisler, G., Beck, H., Bostock, R., Thomas, C., and Luke, E. 2006.The National Plant Diagnostic Network. Plant Dis. 90:128-136.

46. Stack, J. P., Bostock, R. M., Hammerschmidt, R., Jones, J. B., and Luke, E. 2014.The National Plant Diagnostic Network: Partnering to protect plant systems. Plant Dis. 98:708-715.

47. Stenger, D. C., Hall, J. S., Choi, I. R., and French, R. 1998. Phylogenetic relationships within the family Potyviridae: Wheat streak mosaic virus and Brome streak mosaic virus are not members of the genus Rymovirus. Phytopathology 88:782-787.

48. Streck, N. A., Weiss, A., Xue, Q., and Baenzinger, P. S. 2003. Improving predictions of developmental states in winter wheat: a modified Wang and Engel model. Agric. For. Meteorol. 115:139-150.

49. Tanaka, D. L., Karn, J. F., Liebig, M. A., Kronberg, S. L., and Hanson, J. D. 2005.An integrated approach to crop/livestock systems: Forage and grain production for swath grazing. Renew. Agric. Food Syst. 20:223-231.

50. Tatineni, S., McMechan, A., Wosula, E., Wegulo, S., Graybosch, R., French, R., and Hein, G. 2014. An eriophyid mite-transmitted plant virus contains eight genomic RNA segments with unusual heterogeneity in the nucleocapsid protein. J. Virol. 88:11834-11845.

51. University of Utah. 2012. MesoWest. University of Utah Department of Atmospheric Sciences. Online: http://mesowest.utah.edu

52. USDA-APHIS. 2012. Regulated Plant Pest List. United States Department of Agriculture Animal and Plant Health Inspection Service. Online: https://www.aphis.usda.gov/import_export/plants/plant_imports/downloads/RegulatedPestList.pdf

53. USDA-NASS. 2014. United States Department of Agriculture National Agricultural Statistics Service. Online: http://www.nass.usda.gov/

54. Wang, E., and Engel, T. 1998. Simulation of phenological developmment of wheat crops. Agric. Syst. 58:1-24.

55. Yuen, J. E., and Hughes, G. 2002. Bayesian analysis of plant disease prediction. Plant Pathol. 51:407-412.

56. Zadoks, J. C., Chang, T. T., and Konzak, C. F. 1974. A decimal code for the growth stages of cereals. Weed Res. 14:415-421.

Chapter 10

The Trentepohliales (Ulvophyceae, Chlorophyta): An Unusual Algal Order and its Novel Plant Pathogen—*Cephaleuros*

Fred Brooks[1], Fabio Rindi[2], Yasuo Suto[3], Shuji Ohtani[4], and Mark Green[5]

[1] Plant and Environmental Protection Sciences, University of Hawaii at Manoa, Honolulu 96822

[2] Dipartimento di Scienze della Vita e dell'Ambiente, Università Politecnica delle Marche, Ancona 60131, Italy

[3] 5-11-46, Agenogi, Matsue, Shimane 690-0015, Japan;

[4] Faculty of Education, Shimane University, Nishikawatsu, Matsue, Shimane 690-8504, Japan

[5] Department of Biological Sciences, Phycolab Visiting Scholar, The University of Alabama, Tuscaloosa 35487-0345

ABSTRACT

Most plant pathologists know certain algae can be used as gelling agents in culture media. Pathologists practicing in tropical or subtropical environments also know that some algae damage plants. The five genera in the order Trentepohliales (Chlorophyta) are unique and fascinating. Among other characteristics, they are subaerial, bright orange to red in color, and one genus, *Cephaleuros*, is a plant pathogen while another, *Stomatochroon*, is a space parasite. *Cephaleuros* causes algal spot and includes 17 accepted species. Of these, 13 develop between the cuticle and the epidermis of their hosts and four grow intercellularly. The latter are especially damaging, causing chlorosis and branch dieback. Zoospores and gametes germinate on plant surfaces during the rainy season and probably penetrate through breaks in the host cuticle. Their filamentous growth forms thalli that produce sporangiophores and spherical gametangia the following year. Several species of *Cephaleuros* have a broad host range and though their damage is usually superficial, it can be economically important on certain crops. Plant stress is the greatest predisposing factor to this algal disease. Management includes providing plants with sufficient moisture and nutrients, modifying cultural and harvesting practices, and planting resistant cultivars when available.

1. ALGAE AS PLANT PATHOGENS

1.1 Traditional plant pathogens

Fungi, fungus-like stramenopiles, viruses, bacteria, and nematodes cause most diseases of vascular plants. They are responsible for stem rust of wheat, late blight of potato, cucumber mosaic, fire blight, and crown gall. These diseases are well known to plant pathologists and familiar to many others. Organisms that cause pathological effects such as algal spot, however, are less common, usually do not cause severe damage, and are not as well known. The aim of this article is to introduce readers to the fascinating order Trentepohliales and in particular to *Cephaleuros* Kunze in Fries 1829, a trentepohlialean genus well known as a unique algal plant pathogen.

1.2 The algae

The term "algae" includes a broad, diverse assemblage encompassing all organisms that perform oxygenic photosynthesis other than embryophyte land plants (Cavalier-Smith 2007). It is now well established that algae are an unnatural (i.e., nonmonophyletic) group, formed by many evolutionary lineages with separate origins. This heterogeneity is reflected in the huge diversity observed in their size, shape, color, life history, cell wall and chloroplast structure, and secondary metabolites. Nevertheless, there is a tendency to treat algae as a single group because they occur mostly in aquatic habitats, have a simple body structure called a "thallus," and in most cases an autotrophic lifestyle (see glossary).

Reading the term "algae," most people will think about forests of giant kelp (*Macrocystis pyrifera*) that grow along the shores of California and Oregon, or the dense mats of filamentous green algae (e.g., *Cladophora* and *Spirogyra*) found in many streams and lakes. In the course of their evolution, however, several lineages of algae have been able to colonize subaerial (nonaquatic, not usually in or on soil) habitats and survive in conditions of complete desiccation (Lewis and Lewis 2005). Because of their small dimensions and habit, subaerial algae easily can be confused with mosses, lichens, or fungi (Box 1, Fig. 1). Nevertheless, algae are present in most ecosystems and represent an important component of the microbial flora in many nonaquatic sites.

Box 1. Algal parasites misidentified

Species of the plant-parasitic alga *Cephaleuros* are often misidentified. At a glance, the fuzzy, reddish orange growths look like a rust disease (Fig. 1). As a result, leaf and stem infections on tea plantations over a century ago were misleadingly named "red rust" (Mann and Hutchinson 1907). In 1965, a fossilized organism from about 40 million years ago was reported as a fungus, *Pelicothallus villosus*, but later recognized as a species of *Cephaleuros* (Reynolds and Dunn 1984). These algae have been mistaken for coffee rust, powdery mildew, various leaf spots, and bacterial canker (Thompson and Wujek 1997; Wellman 1965).

A closer look at *Cephaleuros* species through a microscope reveals fruiting structures that resemble an unusual downy mildew or other sporangium-forming organism. In 1982, *Veralucia brasiliensis* was described as a fungus using mycological terms, but was later recognized as *Cephaleuros parasiticus* by the same authors (Reynolds and Dunn 1982). Embarrassingly, five mycologist colleagues and two taxonomists reviewed the earlier paper before publication and did not recognize the authors' error. In the southeastern United States, the causal agent of a common leaf spot of*Magnolia grandiflora* previously published as *Glomerella cingulata* in 1947 was identified as *C. virescens* in 1984 (Holcomb and Henk 1984).

Hennen and Figueiredo (1981) tell the story of *Dasyspora gregaria*, a tropical rust first collected in Surinam in 1827. Teliospores and spermagonia were reported on the original material, as was the presence of a powdery mildew, *Oidium parasiticum*. Various collectors have since mistakenly described this powdery mildew as the aecia or uredinia of the *Dasyspora* rust. Hennen and Figueiredo collected fresh material from Brazil in 1979 and had these asexual structures examined. They were not aecia, uredinia, or even powdery mildew, but a new species of the trentepohlialean alga*Stomatochroon*, which lives in the substomatal chambers of its host.

Cephaleuros species are especially difficult to recognize if they are lichenized by one of several fungi in the Ascomycota (Mann and Hutchinson 1907). Their appearance is typical of lichens but they are not usually associated with plant damage. Evidence suggests, however, that the algae cause damage to the plant before they are lichenized, not after (Brooks 2004; Schubert 1981). The presence of fruiting bodies and spores of the fungal component of the lichen may further hinder recognition of the algae.

Figure 1. Algal spot on avocado (*Persea americana*) caused by*Cephaleuros virescens*. Commonly called "red rust," identifying *Cephaleuros* species has confused observers for decades. Even microscopic examination of the lesions reveals hyphae-like filaments, setae, sporangiophores, and swimming spores similar to fungi and the oomycetes.

2. THE ORDER TRENTEPOHLIALES

2.1 Similarities and differences

The order Trentepohliales belongs to the lineage of green algae identified as phylum Chlorophyta. This phylum is sister to the Streptophyta, the lineage that includes some green algae and all land plants (Leliaert et al. 2012). However, placement of this order at class level was a major taxonomic controversy until recently. This uncertainty arose because of a unique combination of morphological and ultrastructural features that make the Trentepohliales one of the most mysterious groups in the plant kingdom. These features include a characteristic type of sporangium (Thompson and Wujek 1997), absence of pyrenoids in the chloroplasts (Van den Hoek et al. 1995), a unique flagellar apparatus (Chapman and Henk 1985), cytokinesis by production of a phragmoplast similar to that of the vascular plants (Chapman and Henk 1986), and plasmodesmata in the cell walls (Chapman and Good 1978). It was only after an analysis of molecular data that placement of the Trentepohliales in the class Ulvophyceae was largely accepted (López-Bautista and Chapman 2003). However, the presence of a phragmoplast-mediated cell division in this group remains a mystery because this mode of cell division is otherwise found only in the Streptophyta.

At first approach, this placement is surprising because the Ulvophyceae are green seaweeds and therefore a strictly marine group: some examples are the sea lettuce *Ulva*and the Atlantic green sponge fingers *Codium fragile*. Even the color is strikingly different. Instead of green, the Trentepohliales often appear orange, yellow, or red because of the accumulation of carotenoid pigments in their cells (López-Bautista et al. 2002). The amount of carotenoids produced depends on environmental conditions and is highest when these algae are exposed to strong light and subjected to a shortage of nutrients (Abe et al. 1999; Hametner et al. 2014a; Ho et al. 1983). Conversely, in shaded habitats the green color of their chloroplasts may not be obscured by carotenoids like β-carotene or astaxanthin (Thompson and Wujek 1997), and some of the genera look nearly identical to filamentous green seaweeds, such as species of *Cladophora*.

2.2 Subaerial algae

The Trentepohliales is one of the most widespread and distinctive groups of subaerial algae. Their center of diversity and abundance is in tropical and subtropical regions, but they can be found in cooler regions where the climate, or local microclimate, is sufficiently humid (Suto and Ohtani 2009). From an ecological point of view the Trentepohliales occupy a wide range of substrata: rocks, tree bark, leaves, stems, fruits, decaying wood, and occasionally bare soil. They also grow on and damage artificial surfaces such as concrete walls, tombstones, marble statues, and plastic (Fig. 2). In certain humid climates, such as western Ireland and Singapore, species of *Trentepohlia* Martius 1817 produce unsightly red streaks on the walls of many houses, requiring periodic cleaning or repainting (Ong et al. 1992; Rindi and Guiry 2002). One of the most peculiar characteristics is that this order includes two genera, *Cephaleuros* and *Stomatochroon* Palm 1934, which live respectively as pathogens and space parasites of vascular plants (Chapman and Good 1983).

2.3 General morphology

Morphologically, the Trentepohliales consist of uniseriate filaments (i.e., filaments formed by a single row of cells), with a variable branching pattern. The way these filaments are arranged is the main character used for distinction at genus level. In some genera the filaments remain separated and form small bushes or prostrate systems with an open branching pattern. In other genera they are united to produce compact disks or pseudoparenchymatous masses (Guiry and Guiry 2014; Printz 1939; Thompson and Wujek 1997). A morphological feature diagnostic of this order is the sporangiate-lateral (Fig. 3), consisting of a zoosporangium and a highly modified branch, the "suffultory" cell, which supports it (Thompson and Wujek 1997).

2.4 Life history

Evidence suggests that the normal life history of three genera in the Trentepohliales,*Trentepohlia*, *Phycopeltis* Millardet 1870, and *Printzina* Thompson and Wujek 1992, is an isomorphic alternation of two generations: a diploid sporophyte and a haploid gametophyte. The other two genera, *Cephaleuros* and *Stomatochroon*, have a heteromorphic alternation of generations. However, it is well documented that the gametophyte thallus can produce both gametangia and zoosporangia, in which case the zoospores reproduce the same generation. There are some reports that biflagellate "swarmers," or presumptive gametes, may germinate and reproduce new thalli without mating, therefore behaving as zoospores instead of gametes (Rindi and Guiry 2002; Suto and Ohtani 2013).

Figure 2. Genera of the Trentepohliales are widely distributed and grow on living and nonliving substrates. **A,** Sinnamary, French Guiana: Species of *Trentepohlia*stain numerous porous surfaces reddish orange. **B,** American Samoa: *Trentepohlia* even lives on nonporous plastics. **C,** Devil's Punchbowl, Oregon:*Trentepohlia flava* growing on the inner rim of a rocky bowl carved by the sea. **D,** Moycullen, Ireland:*Phycopeltis* is a common epiphyte, often with beautifully symmetrical thalli. **E,** Heavy yellowish epiphytic growth of *Printzina lagenifera*. **F,** American Samoa: *Cephaleuros virescens* is one of several species that cause plant damage, like this algal spot of guava (*Psidium guajava*) fruit.

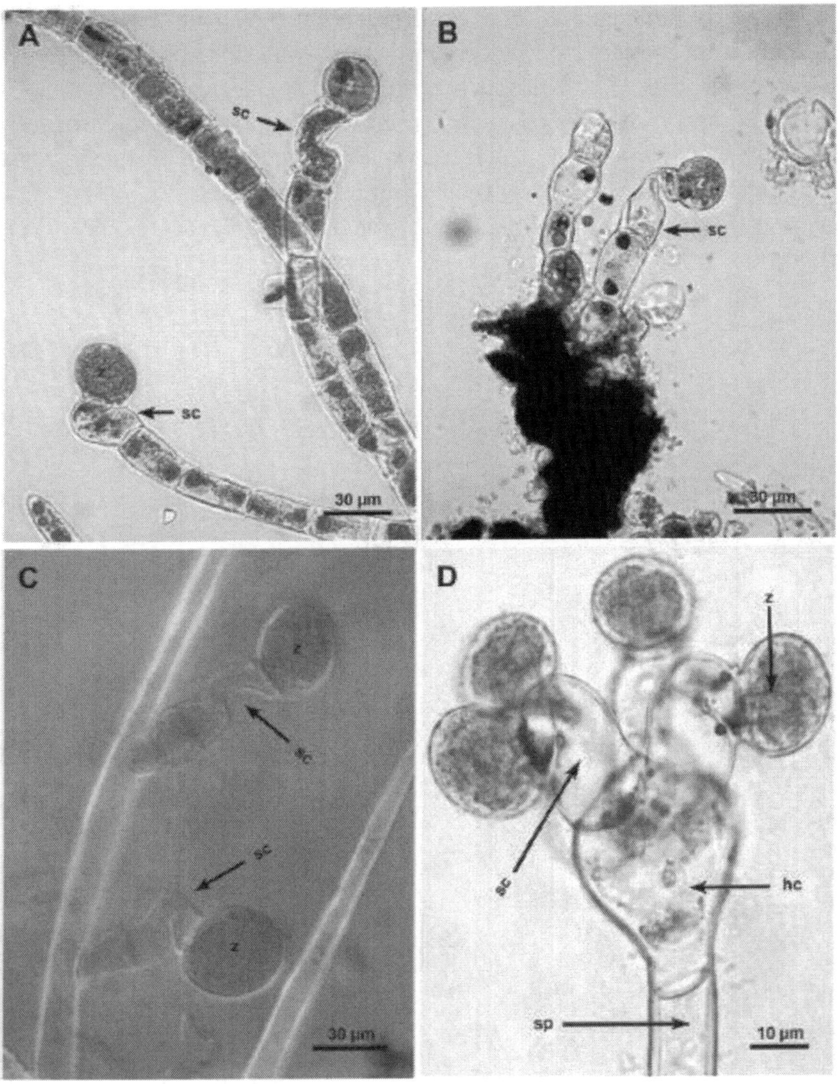

Figure 3. The sporangiate-lateral is unique to the Trentepohliales. It consists of a zoosporangium (z) borne at the end of a crooked "suffultory" cell (sc). When moisture is present the cell straightens, holding the sporangium away from the sporangiophore (sp) and head cell (hc) and promoting windborne dispersal. **A,** *Trentepohlia* sp.; **B,** *Printzina* sp.;**C,** *Cephaleuros minimus*; **D,** *C. virescens*.

Glossary

Biflagellate swarmer—a cell bearing two flagella and capable of active movement even in small volumes of water. In the Trentepohliales, biflagellate swarmers are putative gametes released from a gametangium that do not fuse to form a zygote, but germinate directly and form a new thallus.

Filament—a single strand of cells that forms an algal thallus and its structures; similar to a fungal hypha.

Gametangium—a structure in which gametes are produced.

Gametophyte—one of the two generations of the life cycle typical of most algae and plants. It consists of a haploid thallus that produces gametes and originates from a meiozoospore released from a meiosporangium.

Meiosporangiate-lateral—a branch formed by the suffultory cell and attached meiosporangium; produced on the diploid sporophyte.

Meiosporangium—a sporangium formed on the sporophyte plant in which meiosis occurs.

Meiozoospore (meiospore)—a haploid spore with four flagella produced in a meiosporangium following meiosis.

Mycobiont—the fungal component of a lichen.

Phragmoplast—a structure produced during cell division, consisting of microtubules and small vesicles that fuse to form a cell plate; site of primary cell wall growth typical of plants and some groups of green algae.

Phycobiont—the algal component of a lichen.

Plasmodesmata—microscopic channels that transverse cell walls and link the symplasts (cytoplasm) of adjacent cells.

Pseudoparenchyma—tissue-like structure formed by laterally compacted, short-celled filaments resembling the parenchyma of higher plants.

Pyrenoid—a small body localized in the chloroplasts of many algae, in which intense fixation and concentration of carbon dioxide and accumulation of starch takes place.

Ramulus (pl. ramuli)—a lobe or branch of a thallus formed by laterally joined filaments

Rhizoid—a small appendage formed by one of more cells with the primary function of attachment to a substrate. In *Cephaleuros* it is a reduced, irregular cell growth from the underside of a thallus that runs beneath the host cuticle or between host cells and absorbs water and nutrients.

Space parasite—an endophyte; an alga that may only use its host plant for shelter.

Sporangiate-lateral—a short terminal or lateral branch consisting of a suffultory cell and a zoosporangium.

Sporangium—a structure in which spores are produced.

Sporophyte—one of the two generations of the life cycle typical of most algae and plants. It consists of a diploid thallus that develops from a zygote and produces spores (usually meiozoospores) in meiosporangia.

Subaerial—growing above, or occasionally on, the soil; not subaqueous.

Suffultory cell—a crooked cell to which a sporangium is attached.

Synaptonemal complex—a ladder-like protein complex that binds each homologous pair of chromosomes together and facilitates crossover during meiosis (prophase).

Thallus—the undifferentiated vegetative body of an alga, fungus, or lichen.

Trentepohlialean (adj.)—belonging to or relating to the order Trentepohliales.

Ulvophycean (adj.)—belonging to or relating to the class Ulvophyceae.

3. TAXONOMY

3.1 Genera and species

Some species of Trentepohliales are among the earliest described organisms in the history of systematics. Linnaeus described *Byssus iolithus* (currently *Trentepohlia iolithus*) in 1753 and *B. aureus* (currently *T. aurea*)

in 1759. This order has therefore been studied for a long time and its classification has undergone numerous rearrangements. In the current scheme (Guiry and Guiry 2014), it includes the single family Trentepohliaceae and five genera: *Cephaleuros* (17 species), *Phycopeltis* (25 species), *Printzina* (8 species), *Stomatochroon* (4 species), and *Trentepohlia* (48 species). The taxonomic validity of a sixth genus, *Physolinum* Printz 1920, is controversial and in recent literature this genus has not been separated from *Trentepohlia*.

Here we briefly present the main characteristics of each genus except *Cephaleuros*, which is discussed in detail in a following section.

1) *Trentepohlia.* This genus consists of erect branched filaments, variously developed and arranged, arising from a prostrate system. Sometimes these filaments are compacted to form a crust-like structure. *Trentepohlia* occurs on rocks, tree bark, leaves, soil, and artificial substrata, mainly in exposed habitats (Fig. 2A and B). Bright red patches formed by *Trentepohlia flava* are a common sight on rocks facing the sea in many coastal areas of northern California and Oregon (Fig. 2C). This genus of the Trentepohliales has the widest geographical distribution and though most abundant in the tropics, species of *Trentepohlia* extend into temperate areas (Hametner et al. 2014a) and regions with cold climates such as Alaska, Norway, and Tierra del Fuego.

2) *Phycopeltis.* The thallus of *Phycopeltis* is formed by numerous prostrate, branched filaments that are usually joined to form a closed disk. In some species the thallus has a regular circular outline; in others it is irregular and formed by several lobes, called "ramuli." Initially the alga forms a single layer of cells, and then becomes multilayered in some parts of the thallus, especially where gametangia or zoosporangia are produced. *Phycopeltis* is largely an epiphyllous alga, occurring above the cuticle on the upper leaf surface of numerous vascular plants (Fig. 2D). Its thalli are occasionally found on other parts of plants—stems, branches, twigs, and fruits—and on inert surfaces. Partially shaded sites with constantly moist conditions, especially at the margins of tropical and temperate rainforests, are the most common habitat of *Phycopeltis.*

3) *Printzina.* This genus described by Thompson and Wujek in 1992 includes nine taxa previously belonging to *Trentepohlia.* The reclassification was based mainly on differences in shape of

zoosporangia, arrangement of the sporangiate-laterals, and development of the prostrate parts of the thallus. *Printzina* is usually associated with shaded habitats, but species such as *P. bosseae* and *P. lagenifera* (Fig. 2E) are often found in exposed areas.

4) *Stomatochroon.* Hard to see because it is mostly in the leaf of its host, *Stomatochroon* is the least known and most rarely recorded genus of Trentepohliales. It is an endophytic alga found in the substomatal chambers of vascular plants. For this reason it has been considered a space parasite and may not produce symptoms on its host (Liu and Hu 2013; Zhu et al. 2014). However, severe infections have been associated with water soaking, chlorosis, and necrosis, possibly due to the inhibition of gas exchange from blocked stomata (Brooks 2004; Chapman and Good 1983; Thompson and Wujek 1997; Timpano and Pearlmutter 1983). The thallus of *Stomatochroon* is either a single large cell anchored in a substomatal chamber or a system of branched filaments spreading through the spongy mesophyll. In both cases, only gametangia or single-celled sporangiophores with sporangiate-laterals emerge from the stomata. This genus is probably widespread in tropical regions, but its exact distribution is not easy to assess because it is so difficult to detect.

4. PHYLOGENY

4.1 A puzzling order

The phylogeny of the Trentepohliales has long been one of the least understood aspects of this group. How were the Trentepohliales related to other groups of the previously mentioned green algae and what were the relationships at genus and species level within this order? The unusual combination of features listed in an earlier section as typical of the Trentepohliales suggested possible affinities with four different classes of green algae. These classes were the Charophyceae, Ulvophyceae, Chlorophyceae, and Trebouxiophyceae, with the first two classes generally considered as the main candidates (López-Bautista et al. 2002). As for many other algal groups, substantial advancements in our understanding became possible only with the availability of molecular data.

4.2 An answer

The first molecular data produced for the Trentepohliales were partial SSU rDNA sequences published by Zechman et al. (1990) for two cultured strains, *Trentepohlia* sp. UTEX 1227 and *Cephaleuros virescens* BR-50. López-Bautista and Chapman (2003) presented more complete SSU sequences for eight members of this order: strains of*Cephaleuros, Trentepohlia, Phycopeltis,* and *Physolinum*. The results of this study were the first solid piece of molecular evidence in support of the ulvophycean nature of the order. The analyses of López-Bautista and Chapman (2003) placed the Trentepohliales with moderate to high support in the Ulvophyceae, in an intermediate position between the basal orders Ulotrichales and Ulvales and the derived orders Cladophorales, Bryopsidales, and the Dasycladales. Cocquyt et al. (2010) confirmed this placement in a more recent study based on limited taxon sampling, but on a higher number of molecular markers. Their study included sequences SSU, *rbc*L, and OEE1 mRNA of *T. aurea* in a concatenated dataset of seven markers, recovering this species with moderate support as a sister taxon to the clade formed by Cladophorales, Bryopsidales, and Dasycladales.

4.3 Genera and species

Phylogenetic relationships at genus and species level in the Trentepohliales have been examined in several studies based on SSU and *rbc*L sequence data (López-Bautista et al. 2006; Nelsen et al. 2011; Rindi et al. 2009; Zhu et al. 2014). Unfortunately the assessment is incomplete at present because of limitations in taxon sampling. High-quality sequences are usually obtained from axenic cultures and most trentepohlialean algae grow slowly and are difficult to isolate without contaminants. Consequently for some genera, *Phycopeltis* in particular, the number of sequences available is still limited. Even for *Trentepohlia* and*Printzina*, the genera from which the best datasets are available, the number of sequences reliably identified at species level is a minority. Nevertheless, data acquired in recent years have greatly advanced our knowledge of the phylogeny of this group, revealing a scenario in contrast with the traditional morphological classification (López-Bautista et al. 2006; Nelsen et al. 2011; Rindi et al. 2009; Zhu et al. 2014).

4.4 Five lineages

Studies based on the largest taxon sampling (Nelsen et al. 2011; Rindi et al. 2009) defined five main lineages as well-supported clades, but none correspond to a trentepohlialean genus as defined on morphological grounds (Fig. 4). A large group, clade 1 (Nelsen et al. 2011; Rindi et al. 2009), appears to be the most basal lineage and is sister to all other Trentepohliales. This lineage is formed by species of *Trentepohlia* and *Printzina* with different morphologies and widespread geographical distributions (*Printzina bosseae, T. flava, T. iolithus, T. umbrina*), as well as a number of unidentified strains from lichens sequenced by Nelsen et al. (2011). Among the remaining taxa, a small, highly supported lineage, clade 4 (Nelsen et al. 2011), is formed by unidentified algae isolated from tropical lichens in the Old and New Worlds.

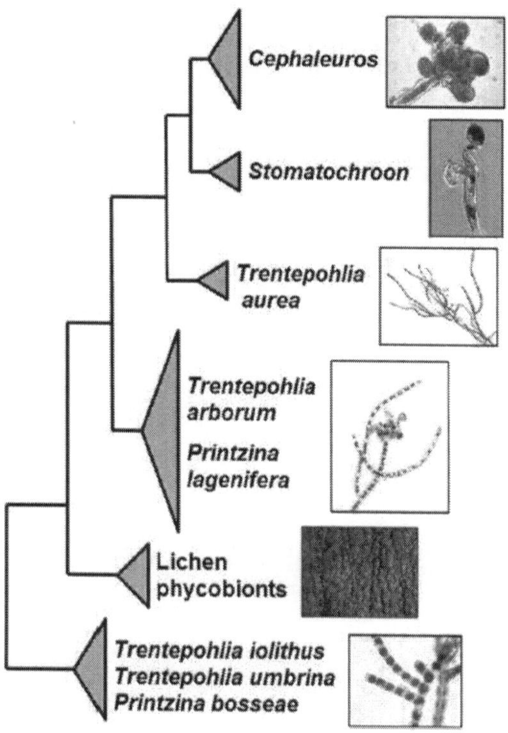

Figure 4. Summary of phylogenetic relationships within the Trentepohliales based on molecular data from López-Bautista et al. 2006, Nelsen et al. 2011, Rindi et al. 2009, and Zhu et al. 2014. *Stomatochroon* photo courtesy of G.-X. Liu.

Sister to this lineage is a large, well-supported clade in which all other Trentepohliales are grouped. Several lineages are recovered within this clade. A large group, clade 2 (Nelsen et al. 2011; Rindi et al. 2009), contains many strains of *Trentepohlia* and *Printzina* from both tropical and temperate regions clustered with numerous lichen phycobionts (i.e., the algal component of a lichen). In this clade are two of the most common trentepohlialean species in tropical regions: *T. arborum* and *Printzina lagenifera*, the type species of*Printzina*. Notably, however, neither of these species represents a monophyletic entity.

4.5 Three taxa represent the most derived lineages

T. aurea (the type species of *Trentepohlia*), *Cephaleuros*, and *Stomatochroon*. In the studies of Rindi et al. (2009) and Nelsen et al. (2011), strains of *T. aurea* formed a well-supported clade closely related to the *Cephaleuros* lineage and the *T. arborum–P. lagenifera* lineages, but its exact placement remained unsupported. Strains of *Cephaleuros*formed a monophyletic group, although with limited support and with a species of*Trentepohlia*, *T. dialepta*, included in it (Nelsen et al. 2011; Rindi et al. 2009). Recently,Zhu et al. (2014) provided the first sequence data for *Stomatochroon*, recovering it as a sister taxon to *Cephaleuros*. This is in agreement with the idea that the endophytic habits of *Cephaleuros* and *Stomatochroon* are derived characters. At present, the phylogenetic position of *Phycopeltis* is still waiting for a detailed assessment, as the relatively few SSU sequences available for this genus have only recently been released.

5. LICHENIZED TRENTEPOHLIALES

5.1 Symbiosis

Trentepohlialean algae often exist in symbiosis with lichen-forming fungi and are one of the most common algal components of tropical and subtropical lichens (Brodo et al. 2001). Lichens are often grouped informally with the cryptogams, plants that reproduce by spores rather than seeds, but this is only partially true. Lichens are not plants; they are composite organisms consisting of a fungus, the *mycobiont*, and at least one alga, the*phycobiont* (Brodo et al. 2001; Chapman and Waters 2002; Grube et al.

2009). Lichens commonly reproduce when fragments of fungal and algal tissue are dispersed and grow to form a new lichen thallus (Brodo et al. 2001).

5.2 The mycobiont

Lichens are classified according to the mycobiont, most of which are in the Ascomycota, or "sac" fungi, with only a few known Basidiomycota, or "club" fungi (Brodo et al. 2001). The fungal component of trentepohlialean lichens is grouped in four distinct classes of the Ascomycota: Arthoniomycetes, Dothideomycetes, Eurotiomycetes, and Lecanoromycetes (Nelsen et al. 2011; Schoch et al. 2009). These fungi are not monophyletic or represented by a single ancestor (DePriest 2004; Gargas et al. 1995). Most can reproduce sexually by ascospores and asexually by conidia. To be successful, however, the spores must come into contact with and colonize a compatible phycobiont and form a new lichen thallus.

5.3 The phycobiont

Ahmadjian (1993) estimated that worldwide 31% of lichen-forming fungi associate with trentepohlialean algae, though Nelsen et al. (2011) suggest this number is closer to 23%. These estimates, however, do not reflect the geographical distribution of trentepohlialean–fungal associations. Research indicates that the Trentepohliales are more prevalent in tropical regions. Therefore a relatively greater number of tropical lichens have trentepohlialean phycobionts, about 45% (Grube et al. 2009; Nelsen et al. 2011; Rindi et al. 2009), and they are especially frequent among lichens that grow on leaves (Ahmadjian 1993; Brodo et al. 2001; Nelsen et al. 2011). Unlike their mycobiont partners, once trentepohlialean algae are colonized by fungi they can no longer form reproductive structures (Hametner et al. 2014a). Recently Hametner et al. (2014a, 2014b) studied the identity of several lichen phycobionts using molecular data. Their results showed a considerable taxonomic diversity, indicating that *Trentepohlia abienta, T. aurea, T. arborum*, and *P. lagenifera* were the morphospecies to which these phycobionts were most closely related.

5.4 Cephaleuros: A victim?

Lichens with *Cephaleuros* as the phycobiont are interesting because the algae are at the very least spatial parasites of their host. Most *Cephaleuros* species are obligate parasites that grow subcuticularly or intercellularly and cause damage to their host plants (Chapman 1976; Green 2012). The symbiotic relationship between the phycobiont and the mycobiont is often unclear, but it displays varying degrees of parasitism (Brooks 2004; Chapman 1976; Mann and Hutchinson 1907; Schubert 1981; Suto and Ohtani 2009, 2011a). For example, *Strigula smaragdula* (Lücking 2008) is an obligate leaf-inhabiting lichen (Fig. 5) and its mycobiont aggressively acquires several trentepohlialean phycobionts (Lücking 1999). The fungus first entangles itself among the algal filaments and assumes the same basic form as the phycobiont. It then uses its haustoria to extract water and nutrients from the algal cells. This process differs from plant-parasitic fungi, however, whose haustoria penetrate plant cell walls. An ultrastructural study of *S. complanata* (Matthews et al. 1989) showed that the cell walls of the phycobiont are able to expand and invaginate as haustoria of the mycobiont invade. Eventually, however, the algal cell wall appears to disintegrate and the cell dies (Matthews et al. 1989). In the *Strigula* species studied, therefore, the mycobiont can be described as a hyperparasite (Chapman 1976; Chapman and Good 1983; Thompson and Wujek 1997).

5.5 Ecosystem change and survival

Lichens undoubtedly play a role in establishing various trentepohlialean genera within natural ecosystems (Green 2012). Relatively little is known, however, about algae of the tropical regions. According to Rindi et al. (2006), "This is particularly worrying both because these are among the most endangered ecosystems on the planet and because recent investigations have shown that many new taxa, including some major evolutionary surprises, can be expected to be discovered in these environments." The microbial communities of tropical and subtropical forests represent a potentially rich biological resource for our global community and a greater knowledge of their diversity is urgently needed (Rindi et al. 2006).

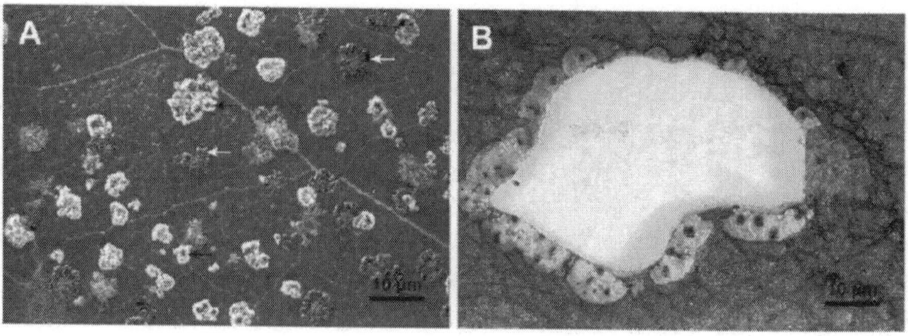

Figure 5. A, The lichen *Strigula smaragdula* in various stages of colonizing the alga (phycobiont) *Cephaleuros virescens*. Black dots (arrows) in the thalli are fruiting bodies of the fungus (mycobiont). Orange thalli are only partially colonized, green thalli fully colonized, and the whitish thalli are deteriorating. Photo courtesy of E. Barnard. **B,** If *Cephaleuros* kills all the plant cells beneath its thallus and the tissue drops from the leaf, it creates a symptom called "shot hole." Here, *S. smaragdula* appears to have lichenized the remaining thallus surrounding the shot hole except for a portion of the alga at the far right.

Recently, climate change has been predicted to significantly affect the frequency and severity of disturbances that shape the world's forests and other natural ecosystems (Bentz et al. 2010). Average mean atmospheric temperature is expected to increase between 1.8 and 4°C by the end of the 21st century (Bentz et al. 2010). It has recently been documented that global warming has influenced the migration of trentepohlialean algae and the tropical lichens that associate with them. A study conducted by Aptroot and van Herk (2007) noted that a majority of the epiphytic lichens rapidly colonizing forests in Western Europe are associated with the Trentepohliales.

5.6 Temperature and disease distribution

Changes in the earth's climate can profoundly affect ecological niches at all levels. According to Erickson et al. (2012), "The widening of the tropical belt has many implications for ecological systems, including the potential expansion of tropical and subtropical diseases, their infectious agents, and vectors that are affected by climatic conditions." In the southern United

States, the Asian tiger mosquito (*Aedes albopictus*) has steadily increased its northern range due to warming temperatures. With this vector comes the threat of various bloodborne diseases such as dengue fever, yellow fever, and chikungunya fever (Erickson et al. 2012). This same scenario emerges with plant diseases. Sudden oak death, caused by an oomycete, has rapidly become entrenched in California's temperate forests and is predicted to increase in severity as spring months continue to become warmer (Pautasso et al. 2012). The incidence and severity of damage associated with *Cephaleuros* are likely to be influenced by changing climate patterns, especially in regions experiencing increased precipitation and warmer temperatures.

6. CEPHALEUROS SPECIES

6.1 History of the genus

Kunze created the genus *Cephaleuros* for specimens of *C. virescens* from Surinam. However, taxonomic confusion about the type species arose because Kunze's specimens had been lost and his description was not sufficiently detailed. Cunningham (1879)described a parasitic alga from India as *Mycoidea parasitica* Cunningham, a new genus and species, but Hariot (1889) pointed out that Cunningham's genus *Mycoidea* was identical to Kunze's genus *Cephaleuros*. Karsten (1891) accepted the older genus name, *Cephaleuros*, added five species, and named *M. parasitica* as *C. mycoidea* Karsten, a new species. Mann and Hutchinson (1907), however, concluded that the reasons given by Karsten were not sufficient to justify the abandonment of *C. virescens*. Printz (1939) reported 13 species in his monograph, recognized the priority of *C. virescens*, and reduced *M. parasitica* and *C. mycoidea* to synonymy. Thompson and Wujek (1997) published a monograph that included 13 species of *Cephaleuros* and one variety and in 2009, Suto and Ohtani added three species from Japan. Detailed histories of the genus have been described by Thompson and Wujek (1997), López-Bautista et al. (2002), and Suto and Ohtani (2009). Guiry and Guiry (2014) currently list 17 species and one variety of the genus (Box 2).

Box 2. Cephaleuros species

List of currently accepted subcuticular and intercellular *Cephaleuros* species (Guiry and Guiry 2014).

Subcuticular species (thallus mainly between host cuticle and epidermis)

Cephaleuros aucubae Suto & Ohtani

Cephaleuros diffusus Thompson & Wujek

Cephaleuros drouetii Thompson & Wujek

Cephaleuros endophyticus (F.E.Fritsch) Printz

Cephaleuros expansa Thompson & Wujek

Cephaleuros henningsii Schmidle

Cephaleuros japonicus Suto & Ohtani

Cephaleuros karstenii Schmidle

Cephaleuros lagerheimii Schmidle

Cephaleuros microcellularis Suto & Ohtani

Cephaleuros solutus Karsten

Cephaleuros tumidae-setae Thompson & Wujek

Cephaleuros virescens Kunze ex E.M.Fries (type species)

Intercellular species (thallus subcuticular, subepidermal, intracellular)

Cephaleuros biolophus Thompson & Wujek

Cephaleuros minimus Karsten

Cephaleuros parasiticus Karsten

Cephaleuros parasiticus var. *nana* Thompson & Wujek

Cephaleuros pilosa Thompson & Wujek

Identifying *Cephaleuros* species requires a close examination of their morphological characteristics. A supportive diagnostic method is counting chromosomes (Box 3, Fig. 6).Suto and Ohtani (2011b) discovered the number of chromosomes among five species of*Cephaleuros* were different yet remained consistent for each species. Current morphological identifications at genus and species levels, however, are being revised by advanced molecular techniques (López-Bautista et al. 2006).

Box 3. Chromosomes

Identifying *Cephaleuros* species using morphological characteristics can be difficult as their form varies with the environment. One approach to this problem was to determine if each species had a unique number of chromosomes. Jose and Chowdary (1977) found different numbers of chromosomes among isolates of both *Cephaleuros solutus* and *C. virescens* and called these isolates cytological races. The presence of these races compromised the effectiveness of this approach. However, Suto and Ohtani (2011b) recently reported stable numbers of haploid chromosomes for each of five species of *Cephaleuros* cultured on agar media. The numbers of chromosomes were as follows: 12 in the filaments of *C. microcellularis*; 22 in the filaments and gametes of *C. aucubae* and 34 in *C. biolophus*; and 18 in the filaments, gametes, and zoospores of *C. japonicus* and 24 in *C. virescens* (Fig. 6). Note: not all cultures produced gametes and zoospores.

7. SYMPTOMS AND SIGNS

7.1 Two growth habits

Species of *Cephaleuros* can be divided into those that grow between the cuticle and epidermis of their host plants (subcuticular) and those that penetrate the epidermis and grow among the cells of their hosts (intercellular) (see Box 2). Most species of*Cephaleuros* grow beneath the cuticle on the upper surface of leaves and on stems and fruits. The few intercellular species, however, grow through the epidermis, palisade, and spongy mesophyll cells and are the most damaging to their hosts.

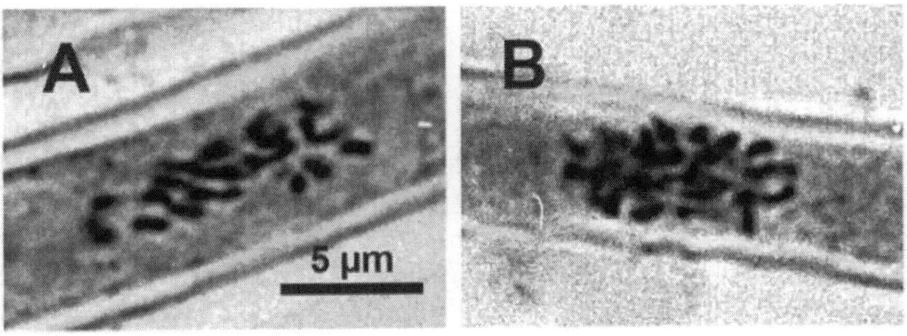

Figure 6. Suto and Ohtani (2011b) consistently reported a correlation between the morphologies of five species of *Cephaleuros* in Japan and the number of chromosomes in gametes, zoospores, or filament cells. Chromosomes in filament cells of **A,***Cephaleuros aucubae* (*n* = 22) and **B,** *C. virescens* (*n*= 24).

7.2 Algal spot

The original name for *Cephaleuros* diseases was "red rust," an apt description of the reddish orange thalli that infect tea (*Camellia sinensis*) in India (Mann and Hutchinson 1907). As mentioned earlier, this name has led to some unfortunate misidentifications (see Box 1). A currently accepted common name for this disease is algal spot, but red rust is still used. If "spot" refers to necrosis, the reaction of the plant to the alga, it is a symptom. If it refers to the thallus of the alga and its projections, then the "spot" is a sign.

7.3 Symptoms

Most subcuticular species of *Cephaleuros* only cause necrosis beneath their thalli (Fig. 7A and B) (Brooks 2004; Ogle 1997; Suto and Ohtani 2009). This necrosis is usually limited to the epidermal cells but occasionally includes palisade parenchyma and can extend to the lower leaf surface (Fig. 7C) (Chapman 1976; Thompson and Wujek 1997). On hosts in Japan, the subcuticular *C. japonicus* produces reddish brown necrosis 3 to 18 mm in diameter beneath thalli on the upper leaf surface, and corresponding yellow discolorations on the lower leaf surface (Suto and Ohtani 2009). Intercellular species cause translaminar necrosis that appears as circular, purplish black spots on upper and lower leaf surfaces.*Cephaleuros biolophus*, for example, destroys the leaf mesophyll and causes cracking of the epidermis of both leaf surfaces (Suto and Ohtani 2009). Lesion diameters of two intercellular species measured 3 to 6 mm for *C. biolophus* and 2 to 25 mm for *C. parasiticus* (Thompson and Wujek 1997). Algal spots on fruits are superficial and usually smaller than on leaves (Joubert and Rijkenberg 1971; Ogle 1997). The most serious plant damage is caused by *Cephaleuros* species that attack leaves and twigs of crops such as tea, citrus, and blackberry (Dewdney 2012; Holcomb et al. 1998; Ramaya et al. 2013;Wolf 1930). Infected twigs and stems are often swollen and cracked (Holcomb et al. 1998). Thalli can coalesce and girdle twigs causing chlorosis, cankers, premature defoliation, and dieback (Joubert and Rijkenberg 1971; Suto and Ohtani 2009; Wolf 1930).

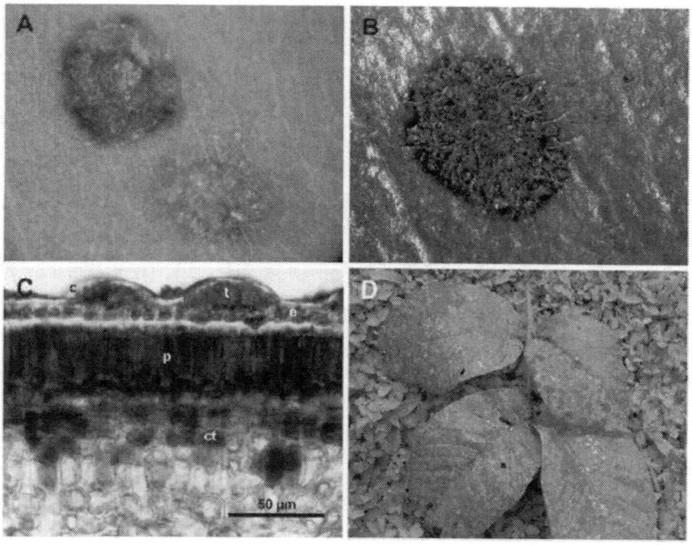

Figure 7. A, Tissue necrosis caused by *Cephaleuros* species is usually limited to cells beneath the thallus. The thallus of *C. virescens* and most of the plant cuticle have been removed from the upper lesion to show necrosis.**B,** On some hosts, *Cephaleuros* causes hyperplasia of the epidermis and palisade parenchyma, raising the thallus above the leaf surface. **C,** An infection of*Camellia japonica* by the subcutaneous alga*Cephaleuros japonicus*: cuticle (c), thallus (t), epidermis (e), palisade parenchyma (p), and corky tissue (ct) formation among spongy mesophyll cells in response to infection. **D,** These yellowish orange, fuzzy thalli on guava leaves are a typical sign of *C. virescens*.

7.4 Signs

The greenish yellow to reddish orange thalli of subcuticular *Cephaleuros* species are quite distinctive (Fig. 7A, B, and D). They are usually slightly raised and have a velvety appearance due to sporangiophores and setae (sterile hairs) growing up through the plant cuticle. Macroscopically, gametangia appear as aggregates or concentric rings of greenish black granules (Suto and Ohtani 2009). The intercellular species are less obvious as their thalli are sometimes scant or below the epidermis, and sporangiophores are usually produced through stomata and the lower epidermis (Chapman and Henk 1985; Suto and Ohtani 2009; Thompson and Wujek 1997).

Subcuticular thalli develop as cells produced by germinated zoospores divide and form branching filaments. As thalli age, the central portion decays, turns whitish, and fresh thallus cells develop around it. In Japan, this condition is called "whitish algal disease" (Suto and Ohtani 2013). The decayed portion of the algal thallus and necrotic leaf tissue below it sometimes drops from the leaves, producing a "shot hole" appearance. When decayed tissue finally spreads to the periphery, the thallus dies (Brooks 2004; Chowdary and Jose 1979).

8. MORPHOLOGY

The thalli of all *Cephaleuros* species, both subcuticular and intercellular, are composed of a prostrate system and an erect system.

8.1 The prostrate system

The filamentous cells of *Cephaleuros* species are cylindrical to irregular in shape and develop by dichotomous or irregular branching of the apical cells (Fig. 8A). If these filaments grow radially and coalesce, they produce an open-ramulate thallus between the cuticle and epidermis (Fig. 8B). If the radial growth of the filaments is compressed laterally, it forms a disk-like, pseudoparenchymatous thallus (Fig. 8C). The margins of this disk can be entire, or lobed with deep sinuses. Intercellular species form loosely branching thalli beneath the cuticle and either above or below the host epidermis (Fig. 8D). Gametangia are spherical to elliptical and formed from terminal cells of the filaments. If the released gametes do not fuse, they form a spherical shape and shed their flagella as they come to rest. Rhizoids form as irregular cellular growth from the underside of the thallus and are thought to anchor the alga in position (Joubert and Rijkenberg 1971). The rhizoids and other cells of the thallus also absorb water and nutrients from the host plant (Ong et al. 1992).

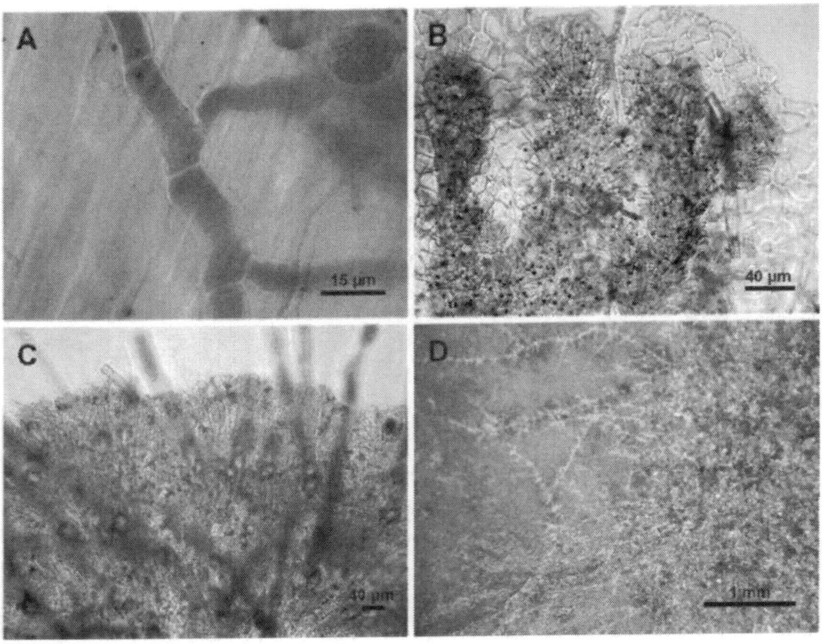

Figure 8. A, Irregular branching of an intercellular filament of*Cephaleuros minimus*. **B,** When individual filaments expand they form branches called "ramuli." This "open-ramulate" thallus of *C. karstenii* has spaces between the ramuli. **C,** The ramuli of *C. virescens*usually coalesce and form an "entire" or disk-like thallus. The blurred projections are multicellular setae. **D,** The filamentous thallus of the intercellular pathogen *C. minimus* is partially visible beneath the upper leaf cuticle.

8.2 The erect system

This system consists of sporangiophores (Fig. 9A–D) and setae (Fig. 9 E–F) that project externally through the cuticle of leaves, twigs, and fruits.

Intercellular species may produce sporangiophores through the upper leaf cuticle but most emerge from stomata or erupt through the lower epidermis. Sporangiophores and setae develop from terminal cells of thalli, usually when one lobe of a branching filament becomes obstructed. Sporangiophores generally consist of 5 to 7 cells and setae from 1 to 33 cells (Thompson and Wujek 1997). Head cells usually form at the apex of sporangiophores and produce sporangiate-laterals (see Fig. 3). There are two exceptions: *C. biolophus* produces head cells and sporangiate-laterals both along sporangiophores and at their apices, and *C. minimus* produces head cells along sporangiophores but with sporangiate-laterals only on one side of sporangiophores (see Fig. 3C and 9D). Instead of terminal head cells, the last few cells of *C. minimus* sporangiophores are tapered and sterile (Brooks 2004; Thompson and Wujek 1997).

9. LIFE CYCLE

9.1 Asexual cycle

Zoospores germinate and develop into thalli. Sporangiophores are produced from terminal thallus cells and develop sporangiate-laterals. When rain or dew wet the zoosporangia they release quadriflagellate zoospores (Fig. 10) as primary inoculum. From 8 to 64 zoospores are produced in each zoosporangium (Chapman and Henk 1985; Guiry and Guiry 2014;Thompson and Wujek 1997). At first, zoospores appear to have two flagella, but each flagellum consists of two flagella twisted together. Zoospores have the same shape as gametes of the same species, but are one-third to one-half larger. Thompson and Wujek (1997) provide an excellent review of the asexual and sexual life cycles of *Cephaleuros*species.

9.2 Sexual cycle

Gametangia develop from terminal and lateral thallus cells. From 8 to 64 biflagellate isogametes (Fig. 10) are produced in the gametangia and are half to two-thirds the size of zoospores of their respective species (Thompson and Wujek 1997). They are spindle shaped when released, becoming ellipsoid to fusiform during their active swimming phase. The gametes fuse outside of gametangia if sufficient moisture is present; otherwise, they fuse within gametangia (Thompson and Wujek 1997).

Zygotes grow into dwarf plants (2n) bearing meiosporangia on meiosporangiate-laterals. Meiosporangia produce quadriflagellate meiozoospores that germinate and produce new thalli (n). The diploid phase of the life cycle is considered to occur only in the dwarf plants and the site of meiosis is purported to be in the meiosporangia. This cycle is a heteromorphic alternation of generations and is repeated as more gametangia mature. Using transmission electron microscopy (TEM), Chapman and Henk (1981) discovered sporangia with multiple nuclei containing synaptonemal complexes in specimens of *C. virescens*. This observation indicated that meiosis had occurred. Based on the proposed life cycle, they concluded that the TEM specimen had contained a dwarf sporophyte growing on the gametophyte thallus. López-Bautista and others (2006) reported that the DNA content in 2C nuclei of *C. parasiticus* gametophytes closely approximated 50% of the 4C values in the sporophyte and was consistent with the presumptive sexual life cycle and alternation of haploid and diploid phases.

9.3 Presumptive gametangia and biflagellate swarmers

Although the above observations support the proposed existence of haploid and diploid generations, Suto and Ohtani (2013) reported no evidence of sexual conjugation of gametes in five species of *Cephaleuros*. They only recorded direct germination and formation of a haploid thallus. Moreover, they did not observe diploid dwarf plants among natural populations of these species at their study sites in Japan, or on three hosts from India (Suto et al. 2014). Rindi and Guiry (2002) reported that in some Trentepohliales from western Ireland, including *Printzina* and *Trentepohlia*, "biflagellate swarmers," which have been called gametes, were liberated from "presumptive gametangia." These swarmers germinated without conjugation and produced new plants. Therefore, Suto and Ohtani (2013) also suggest that in populations of *Cephaleuros*, gametangia be called "presumptive gametangia" and gametes be called "biflagellate swarmers." Since gametes usually fuse near or within their gametangia, it could also be argued that if they mate it would likely be with genetically identical gametes, reducing the evolutionary value of sexual reproduction.

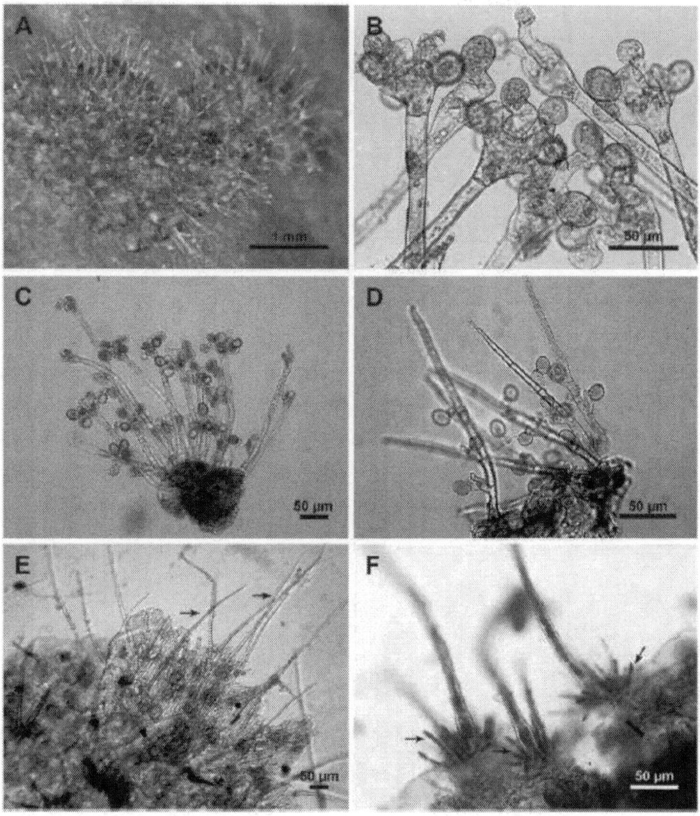

Figure 9. A, Typical lesion on an upper leaf surface caused by*Cephaleuros virescens*. The erect portion of the alga consists of sporangiophores with head cells and sporangiate-laterals, and setae. **B,** Head cells and sporangiate-laterals of *C. japonicus*. **C,** Cluster of *C. parasiticus* sporangiophores emerging from the lower leaf surface. **D,** Sporangiate-laterals of *C. minimus* form on one side of sporangiophores; the pointed, three-celled apex is sterile. **E,** Multicellular setae (arrows) of *C. karstenii*. **F,** Short setae (arrows) of *C. henningsii*.

10. SEASONAL DISEASE CYCLE

10.1 Seasons and development

Most scientists agree that *Cephaleuros* species are plant pathogens. Reports by several researchers on its disease cycle vary, however, and are probably related to differences in climate. Algal spot is a monocyclic disease. In general, dispersal of gametes and zoospores, host infection, and development of thalli and lesions occur during the rainy season (Chowdary and Jose 1979; Mann and Hutchinson 1907; Salleh and Kamsare 1994;Wolf 1930). Reproduction, dispersal, and infection take place the following year (Mann and Hutchinson 1907; Suto and Ohtani 2013; Wolf 1930), with gametangia forming earlier than zoosporangia (Chowdary and Jose 1979; Marche-Marchad 1977; Suto and Ohtani 2013). The seasonal development of *Cephaleuros* has been studied in more detail in parts of Florida and Japan.

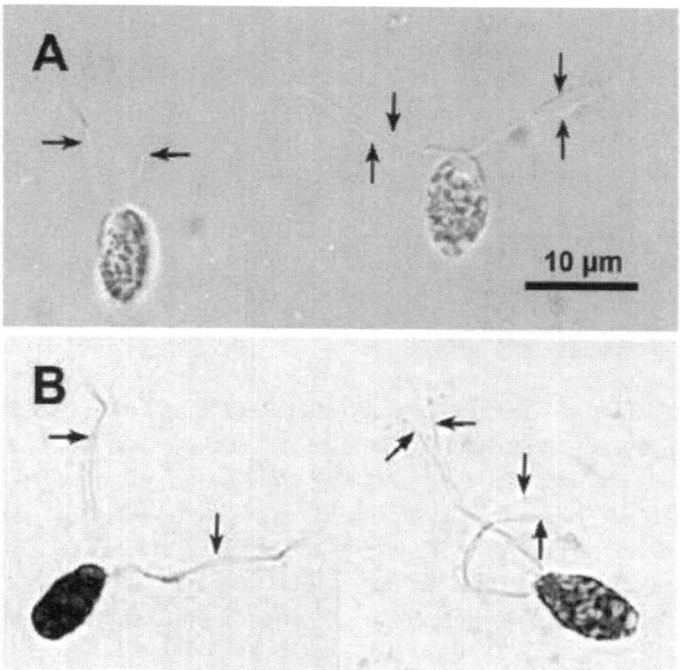

Figure 10. When released from their gametangia and zoosporangia, gametes and zoospores both appear to have two flagella (arrows). Zoospores have four flagella, however, twisted together in two pairs. **A,**Gametes of *C. virescens* with two flagella (left) and zoospores with four flagella (right). **B,** Gametes of *C. japonicus* with two flagella (left) and zoospores with four flagella (right), both stained with Safranin O.

10.2 Subtropical Florida

Wolf (1930) reported similar annual disease cycles on *Magnolia grandiflora* and *Citrus*species in Orlando, Florida. *Magnolia* formed new leaves in May and infections (thalli) were visible by the end of the rainy season in September and October. By January of the next year, setae were protruding above the leaf cuticle. At the beginning of the rainy season in May, sporangiophores were producing zoosporangia and zoospores.

10.3 Temperate Japan

Suto and Ohtani (2013) described the seasonal development of *Cephaleuros* species in a temperate region of Honshu, Japan. They determined that young thalli appeared mainly on 1- and 2-year-old leaves of various woody hosts at the end of the rainy season in June and July. Thalli continued to develop from April through September of the following year, producing gametangia and zoosporangia on 2- and 3-year-old leaves. Gametes of *C. microcellularis* were released from late March to late May and gametes of four other species from late April to late July. Zoospores of *C. biolophus, C. japonicus*, and *C. virescens* were released from mid-May to early August. As in subtropical Florida, one infection cycle was completed each year in temperate Japan.

10.4 Inoculum production

Inoculum is produced the year following infection and disease can be severe (Brooks 2004;Cunningham 1879; Mann and Hutchinson 1907; Marlatt and Campbell 1980; Suto and Ohtani 2013; Wolf 1930). Thompson and Wujek (1997) estimated zoospore production on*Abutilon* species as follows. A 2-mm-diameter thallus of *C. virescens* produced an average of 250 sporangiophores, each with 4 to 11 sporangiate-laterals (zoosporangia). If each zoosporangium contained 8 zoospores, *C. virescens* would produce 8,000 to 22,000 zoospores per 2-mm thallus. With an average of 9 thalli per cm^2 of leaf surface, this alga would produce 72,000 to 198,000 zoospores per cm^2. This is a conservative estimate as the reported number of zoospores per zoosporangium varies. Chapman and Henk (1985)found 8 or 16 zoospores in zoosporangia of *C. parasiticus* and 32 or 64 in *C. virescens*. The number of viable spores is much lower, however, as gametes and zoospores frequently burst a few minutes after their release (Cunningham 1879; Sarma 1986; Suto and Ohtani 2013; Ward 1884).

10.5 Inoculum distribution

Sporangiate-laterals extend from head cells on sporangiophores (see Fig. 3). The supporting branch, the suffultory cell, is wider at the base and tapers apically to a short neck that ends in a spherical or oval zoosporangium. Under moist conditions, the suffultory cell swells and straightens, holding the zoosporangium away from the head cell (López-Bautista et al. 2002). At maturity, the zoosporangium abscises and is carried by the wind. This is the main means of spore dispersal (Chapman and Good 1978) and another unique feature of the Trentepohliales. Rain and moving water, insects, and "jarring or brushing" can also dislodge and move zoosporangia on the plant surface or from plant to plant (Joubert and Rijkenberg 1971; Mann and Hutchinson 1907; Thompson and Wujek 1997).Suto and Ohtani (2013) recorded the release of gametes and zoospores 2 to 5 min after placing pieces of thallus in water. Though most reports indicate only water is needed to release the motile cells, they observed that some form of physical (mechanical) disturbance was required.

10.6 Host penetration

Researchers have yet to determine how *Cephaleuros* species enter their host plants (Thompson and Wujek 1997). There is no evidence of enzymatic or mechanical penetration of the cuticle. Mann and Hutchinson (1907) thought zoospores germinated on the plant surface, formed small primary thalli, and then breached the cuticle by "pushing their way through." Wolf (1930), however, observed that most of these superficial thalli died. Only spores that found breaks in the cuticle and developed thalli beneath it survived. In support of this, Suto and Ohtani (2011c) demonstrated that leaves of certain plants wounded by cutting or rubbing were more susceptible to infection than nonwounded plants.

11. INOCULATION STUDIES AND HOST RANGE

11.1 Early work

Inoculation experiments can be used to determine the pathogenicity and virulence of a suspected causal agent of disease and its host range. However, few inoculation studies with *Cephaleuros* species have been published. In the 1920s, Wolf (1930) detected no host infections following inoculations with zoosporangia, fragments of thalli, or filaments from pure cultures of *C. virescens*. Fifty years later Chowdary and Jose (1979) were also unable to produce infections using zoospores as inoculum.

11.2 Newer work

Suto and Ohtani (2011b) successfully inoculated seedlings of various woody plants in Japan with five species of *Cephaleuros*. The algae were cultured on agar media and then applied directly to leaves, or applied as a suspension of fragmented thalli. Seasonal development of disease in the controlled studies was the same as in natural inoculations. Inoculation experiments with biflagellate swarmers and zoospores, which are natural inocula in the field, should be used in future bioassays.

11.3 Host specificity

The host range of *Cephaleuros* species in the tropics and subtropics is extensive. It includes both angiosperms and gymnosperms, with angiosperms most affected (Brooks 2004; Holcomb 1986; Marlatt and Alfieri 1981; Suto and Ohtani 2009). The most commonly reported species is *C. virescens* (Thompson and Wujek 1997). Batista and Lima (1949)listed *C. virescens* on 448 hosts in the state of Pernambuco, Brazil, some by their Latin binomial and others only by their genus or common names. Holcomb (1986) reported *C. virescens* on 218 host genera, species, and cultivars in Louisiana alone. No other*Cephaleuros* species were listed by Batista and Lima or by Holcomb. Of five *Cephaleuros*species collected from 152 plant taxa in American Samoa, 115 were *C. virescens* (Brooks 2004). In Japan, Suto and Ohtani (2009) collected 45 plant species of which 25 were infected with *C. virescens*.

Though *C. virescens* has a relatively broad host range, other species appear to have a limited number of hosts. Reports of host specificity are usually based on field observations, but controlled field inoculation studies by Suto and Ohtani (2011c) in Japan indicated definite host preferences. The following combinations of algae and host plants produced infections: *Cephaleuros aucubae* on *Aucuba japonica*; *C. biolophus* on *Quercus phillyraeoides*; *C. japonicus* on *Camellia japonica*, *Camellia sinensis*, *Persea thunbergii*, and *Symplocos lucida*; *C. microcellularis* on *P. thunbergii* and *S. lucida*; and *C. virescens*on *Camellia japonica*, *Camellia sinensis*, and *Persea thunbergii*. It should be noted that molecular evidence is already revising morphological identifications of *Cephaleuros* at the species level. For example, a phylogenetic study of *C. virescens* strains from the United States, South Africa, and Taiwan determined they are not a monophyletic group, though similar morphologically (López-Bautista et al. 2006). Ongoing genetic analyses and other approaches will continue to redefine our morphological classification of the genus and the host range of its species.

12. DISEASE IMPACT

12.1 Economics

Leaf infections caused by *Cephaleuros* species usually are not economically important (Dewdney 2012; Holcomb 1986; Joubert and Rijkenberg 1971). Superficial lesions on fruits such as guava, oranges, and smooth-skinned avocados, however, can affect their marketability. Leaf infections may also decrease the aesthetics of natural and ornamental landscapes. Slow-growing lesions on leaves seldom affect plant growth unless severe, with the exception of tea. Since the tea plant is cultivated for its leaves, foliar infection is a special problem and can affect quality and yield (Huq et al. 2010; Ponmurugan et al. 2010; Ramaya et al. 2013).

12.2 India and Japan

C. virescens and *C. parasiticus* have been important pathogens of tea in India since the 1800s (Cunningham 1879; Mann and Hutchinson 1907), with plantations affected in the north and south of the country (Joubert and Rijkenberg 1971; Ponmurugan et al. 2010; Ramaya et al. 2013; Suto and Ohtani 2013). Foliar infection is a source of inoculum and can increase the incidence of the more serious stem infections (Mann and Hutchinson 1907). The intercellular pathogen *C. parasiticus* damages the vascular tissue of young stems, leading to defoliation, dieback, and occasionally plant death. *Cephaleuros* species are commonly observed throughout southwestern Japan. They mainly infect leaves of evergreen trees in gardens, parks, and natural forests, damaging their aesthetic value (Suto and Ogawa 1988). Though algal leaf spot caused by *C. virescens* is widely distributed in tea plantations in Japan, the damage is not serious (Ezuka and Ando 1994).

13. DISEASE MANAGEMENT

13.1 Plant stress

Several authors have reported an inverse relationship between the severity of infection caused by *Cephaleuros* species and host vigor (La Rue 1923; Schubert 1981). Joubert and Rijkenberg (1971) proposed that only healthy plants are able to form the protective suberized tissue that limits algal growth. Plant stress caused by insufficient water or nutrients, poor soil conditions, mechanical or over-harvesting, wounding, and disease are considered major predisposing factors to severe algal infection (Huq et al. 2010; Ramaya et al. 2013; Suto and Ohtani 2011b).

13.2 Cultural control

Reducing plant stress and improving plant vigor is best addressed through cultural practices. These practices include proper irrigation and plant nutrition, thinning to improve aeration in the canopy, proper harvesting techniques, and control of pests and diseases (Huq et al. 2010; Ramaya et al. 2013).

13.3 Managing *Cephaleuros* on tea in India

Most of the research on management of algal infections has been done in India, prompted by heavy losses among the country's tea plantations. In general, copper-based sprays are the most commonly recommended chemical control method (Joubert and Rijkenberg 1971;Mann and Hutchinson 1907; Ramaya et al. 2013; Wolf 1930). One study determined that fungicides provided better protection than biological control agents, which were more effective than the algicides tested (Ramaya et al. 2013). More specifically, the systemic fungicides carbendazim and hexaconazole provided better disease control than the contact fungicides copper oxychloride and mancozeb. Two species of *Streptomycin* reduced the incidence of infection by *Cephaleuros* better than the bacteria *Pseudomonas* and *Bacillus*, or the fungus *Trichoderma*. Disinfesting pruning tools in various algicidal solutions between plants reduced disease incidence up to 25% and hand-plucking leaves produced fewer infections than either shearing or over-harvesting.

13.4 Resistant cultivars

A recent survey of tea plantations in southern India evaluated the resistance of various clonal tea cultivars to *Cephaleuros parasiticus* (Ponmurugan et al. 2010). A majority of the clones had either a moderate or low susceptibility and four of the most popular clones were described as tolerant. Climate was also a factor in clonal susceptibility. Though these results are promising, breeding for resistance to *Cephaleuros* species appears to be relatively unexplored.

14. CONCLUSIONS AND FUTURE RESEARCH

The peculiar biology of the Trentepohliales presents opportunities for exciting and creative research. Despite recent advancements, the diversity of this order is still poorly understood. There is an urgent need for a modern systematic and taxonomic assessment founded on solid molecular data with wide taxon sampling. Although this is common to many algal groups, it is a mandatory requirement for the Trentepohliales, in which morphology offers relatively few characters and is severely affected by phenotypic plasticity. Accordingly, most in need of assessment is the widely reported paraphyletic species, *C. virescens*. This work should be supplemented with detailed observations of morphological characteristics in different environments, inoculation studies to determine host selectivity, and clarification of the infection process. Furthermore, assessing disease development under various field conditions will improve management options, as will resistance breeding for high-value crops.

The development of next-generation sequencing techniques in recent years has greatly expanded the body of genomic and transcriptomic (all mRNA present in a cell or population of cells) data available for algae. The first full genome sequencing project for a trentepohlialean alga, *Trentepohlia flava*, is in progresshttp://bridge.botany.ubc.ca/herbarium/details.php?db= ubcalgae.fmp12&=ubcalgae_web_details&recid=210232&ass_num=A8924 8. Similarly, the first transcriptomic analysis, for *Trentepohlia iolithus*, was recently published (Li et al. 2014). Analysis of genomes and transcriptomes of the Trentepohliales can provide great insights into their physiology and ecology. Pathogenic members of the order are good subjects for these studies. It would be particularly interesting to compare the transcriptomic profiles of subcuticular *Cephaleuros* species with its more virulent intercellular species. This could lead to a better understanding of plant-microbe interactions and improved crop management strategies.

General information on the ecology, natural history, and distribution of the Trentepohliales is always a valuable contribution. We hope this article will encourage plant pathologists, phycologists, and others based in tropical regions to look for and appreciate these algae.

ACKNOWLEDGMENTS

Our thanks to Russell Chapman for his editing and guidance and to Debra Waters for her helpful review. We are also grateful to Juan López-Bautista and Daniel Wujek for their support. We thank the following for permission to use photos: Fig. 4, Nova Hedwigia 2013, 97(3-4):342(no. 14) www.schweizerbart.de; Fig. 5A, Florida Department of Agriculture and Consumer Services, Bugwood.org; Fig. 9C, Pacific Science 2004, 58(3):423A; and Michael Guiry and Matthew Nelsen.

REFERENCES

1. Abe, K., Nishimura, N., and Hirano, M. 1999. Simultaneous production of beta-carotene, vitamin E and Vitamin C by the aerial microalga Trentepohlia aurea. J. Appl. Phycol. 11:331-336.

2. Ahmadjian, V. 1993. The lichen symbiosis. John Wiley and Sons, Inc., NY.

3. Aptroot, A., and van Herk, C. M. 2007. Further evidence of the effects of global warming on lichens, particularly those with Trentepohlia phycobionts. Environ. Pollut. 146:293-298.

4. Batista, A. C., and Lima, D. A. 1949. Lista de suscetiveis da alga Cephaleuros mycoidea Karst. em Pernambuco (Host list of the alga Cephaleuros mycoidea Karst. in Pernambuco). Boletim Sec. Agr. Indus. Com. Pernambuco 16:32-46.

5. Bentz, B. J., Régnière, J. C., Fettig, J., Hansen, E. M., Hayes, J. L., Hicke, J. A., and Seybold, S. J. 2010. Climate change and bark beetles of the western United States and Canada: direct and indirect effects. Bioscience 60:602-613.
Brodo, I. M., Sharnoff, S. D., and Sharnoff, S. 2001. Lichens of North America. Yale University Press, New Haven, CT.

6. Brooks, F. 2004. Plant-parasitic algae (Chlorophyta: Trentepohliales) in American Samoa. Pac. Sci. 58:419-428.

7. Cavalier-Smith, T. 2007. Evolution and relationships of algae: major branches of the tree of life. Pages 21-55 in: Unravelling the Algae: The Past, Present and Future of Algal Systematics. J. Brodie and J. Lewis, eds. The Syst. Assoc. Spec. Vol. 75. CRC Press,Boca Raton, FL.

8. Chapman, R. L. 1976. Ultrastructural investigation on the foliicolous pyrenocarpous lichen Strigula elegans (Fee). Mull. Arg. Phycologia 15:191-196.

9. Chapman, R. L., and Good, B. H. 1978. Ultrastructure of plasmodesmata and cross walls in Cephaleuros, Phycopeltis and Trentepohlia (Chroolepidaceae; Chlorophyta). Br. Phycol. J. 13:241-246.

10. Chapman, R. L., and Good, B. H. 1983. Subaerial symbiotic green algae: interactions with vascular plant hosts. Pages 173-204 in: Algal Symbiosis: A Continuum of Interaction Strategies. L. J. Goff, ed. Cambridge University Press, Cambridge, UK.

11. Chapman, R. L., and Henk, C. 1981. Observations on the putative dwarf sporophyte ofCephaleuros Kunze (Chlorophyta; Chroolepidaceae). Page 162 in: Proc. 13th Int. Bot. Congr.

12. Chapman, R. L., and Henk, M. C. 1985. Observations on the habit, morphology and ultrastructure of Cephaleuros parasiticus (Chlorophyta) and a comparison with C. virescens. J. Phycol. 21:513-522

13. Chapman, R. L., and Henk, M. C. 1986. Phragmoplasts in cytokinesis of Cephaleuros parasiticus (Chlorophyta) vegetative cells. J. Phycol. 22:83-88.

14. Chapman, R. L., and Waters, D. A. 2002. Lichenization of the Trentepohliales. Pages361-371 in: Symbiosis: Mechanisms and Model Systems. Springer, Dordrecht, The Netherlands.

15. Chowdary, Y. B. K., and Jose, G. 1979. Biology of Cephaleuros Kunze in nature. Phykos 18:1-9.

16. Cocquyt, E., Verbruggen, H., Leliaert, F., and De Clerck, O. 2010. Evolution and cytological diversification of the green seaweeds (Ulvophyceae). Mol. Biol. Evol. 27:2052-2061.

17. Cunningham, D. D. 1879. On a new genus of parasitic algae, and the part which it plays in the formation of certain lichens. Trans. Linn. Soc. Lond. Bot. Ser. 2 1:301-316.

18. DePriest, P. T. 2004. Early molecular investigations of lichen-forming symbionts: 1986-2001. Annu. Rev. Microbiol. 58:273-301.

19. Dewdney, M. 2012. Citrus disease spotlight: Algal disease. Citrus Industry, Page 20. Retrieved February 27 2014

20. Erickson, R. A., Hayhoe, K., Presley, S. M., Allen, L. J. S., Long, K. R., and Cox, S. B. 2012. Potential impacts of climate change on the ecology of dengue and its mosquito vector the Asian tiger mosquito (Aedes albopictus). Environ. Res. Lett. 7:034003

21. Ezuka, A., and Ando, Y. 1994. Diseases of Tea. Japan Plant Protection Association.Tokyo (in Japanese).

22. Gargas, A., DePriest, P. T., Grube, M., and Tehler, A. 1995. Multiple origins of lichen symbioses in fungi suggested by SSU rDNA phylogeny. Science 268:1492-1495.

23. Green, M. 2012. The Trentepohliales (Ulvophyceae, Chlorophyta) from coastal South Carolina [Dissertation]. Tennessee Technological University, Cookeville, TN. Available from ProQuest, UMI Dissertations Publishing.

24. Grube, M., Cardinale, M., de Castro, J. V., Müller, H., and Berg, G. 2009. Species-specific structural and functional diversity of bacterial communities in lichen symbioses. ISME J. 3:1105-1115.

25. Guiry, M. D., and Guiry, G. M. 2014. AlgaeBase. Worldwide electronic publication, National University of Ireland, Galway.

26. Hametner, C., Stocker-Wörgötter, E., and Grube, M. 2014a. New insights into diversity and selectivity of trentepohlialean lichen photobionts from the extratropics. Symbiosis 63:31-40.

27. Hametner, C., Stocker-Wörgötter, E., Rindi, F., and Grube, M. 2014b. Phylogenetic position and morphology of lichenized Trentepohliales (Ulvophyceae, Chlorophyta) from selected species of Graphidaceae. Phycol. Res. 62:170-186.

28. Hariot, M. P. 1889. Note sur le genre Cephaleuros. J. Bot. 3:274-276.

29. Hennen, J. F., and Figueiredo, M. B. 1981. The hyphoid aecium, a rust-alga association (Dasyspora-Stomatochroon), and other corrections to Neotropical rusts (Uredinales). Mycologia 73:350-355.

30. Ho, K. K., Tan, K. H., and Wee, Y. C. 1983. Growth conditions of Trentepohlia odorata(Chlorophyta, Ulotrichales). Phycologia 22:303-308.

31. Holcomb, G. E. 1986. Hosts of the parasitic alga Cephaleuros virescens in Louisiana and new host records for the continental United States. Plant Dis. 70:1080-1083

32. Holcomb, G. E., and Henk, C. 1984. Spot of Magnolia grandiflora. Phytopathology 74:822.

33. Holcomb, G. E., Vann, S. R., and Buckley, J. B. 1998. First report of Cephaleuros virescens in Arkansas and its occurrence on cultivated blackberry in Arkansas and Louisiana. Plant Dis. 82:263.

34. Huq, M., Ali, M., and Islam, M. S. 2010. Efficacy of muriate of potash and foliar spray with fungicides to control red rust disease (Cephaleuros parasiticus) of tea. Bangladesh J. Agric. Res. 35:273-277.

35. Jose, G., and Chowdary, Y. B. K. 1977. Karyological studies on Cephaleuros Kunze. Acta Bot. Indica 5:114-122.

36. Joubert, J. J., and Rijkenberg, F. H. J. 1971. Parasitic green algae. Annu. Rev. Phytopathol. 9:45-64.

37. Karsten, G. 1891. Untersuchungen über die Familie der Chroolepideen. Annu. Jard. Bot. Buitenzorg 10:1-66.

38. La Rue, C. D. 1923. Two unreported parasites of Hevea brasiliensis. Pages 69-71 in: Papers from the Michigan Academy of Science, Arts and Letters.

39. Leliaert, F., Smith, D. R., Moreau, H., Herron, M. D., Verbruggen, H., Delwiche, C. F., and De Clerck, O. 2012. Phylogeny and molecular evolution of the green algae. Crit. Rev. Plant Sci. 31:1-46.

40. Lewis, L. A., and Lewis, P. O. 2005. Unearthing the molecular phylodiversity of desert soil green algae (Chlorophyta). Syst. Biol. 54:936-947.

41. Li, Q. Q., Liu, J. G., Zhang, L. T., and Liu, Q. 2014. De novo transcriptome analysis of an aerial microalga Trentepohlia iolithus: pathway description and gene discovery for carbon fixation and carotenoid biosynthesis. PLoS One 9:e108488.

42. Liu, G.-X., and Hu, Z.-Y. 2013. New observations on a rare endobiotic green alga,Stomatochroon lagerheimii, from a tropical area in China. Nova Hedwigia 97:429-436.

43. López-Bautista, J. M., and Chapman, R. L. 2003. Phylogenetic affinities of the Trentepohliales inferred from small-subunit rDNA. Int. J. Syst. Evol. Microbiol. 53:2099-2106.

44. López-Bautista, J. M., Rindi, F., and Guiry, M. D. 2006. Molecular systematics of the green algal order Trentepohliales: an assessment based on morphological and molecular data. Int. J. Syst. Evol. Microbiol. 56:1709-1715.

45. López-Bautista, J. M., Waters, D. A., and Chapman, R. L. 2002. The Trentepohliales revisited. Constancea 83.

46. Lücking, R. 1999. Ecology of foliicolous lichens at the 'Botarrama' trail (Costa Rica), a Neotropical rainforest. IV. Species associations, their salient features and their dependence on environmental variables. Lichenologist 31:269-289.
Lücking, R. 2008. Foliicolous Lichenized Fungi. Flora Neotropica Monograph 103. Organization for Flora Neotropica and New York Bot. Garden, Bronx, NY.

47. Mann, H. H., and Hutchinson, C. M. 1907. Cephaleuros virescens Kunze: the 'Red rust' of tea. Mem. Dep. Agr. India. Bot. 1:1-33.

48. Marche-Marchad, J. 1977. Observation sur la morphogenèse du Cephaleuros virescensKunze Chlorophycée, Trentepohliale. Bull. I.F.A.N. 39:1-22.

49. Marlatt, R. B., and Alfieri, S. A., Jr. 1981. Hosts of a parasitic alga, CephaleurosKunze, in Florida. Plant Dis. 65:520-522.

50. Marlatt, R. B., and Campbell, C. W. 1980. Susceptibility of Psidium guajava selections to injury by Cephaleuros sp. Plant Dis. 64:1010-1011.

51. Matthews, S., Tucker, S., and Chapman, R. 1989. Ultrastructural features of mycobionts and trentepohliaceous phycobionts in selected subtropical crustose lichens. Bot. Gaz. 150:417-438.

52. Nelsen, M. P., Rivas Plata, E., Andrew, C. J., Lücking, R., and Lumbsch, T. 2011.Phylogenetic diversity of trentepohlialean algae associated with lichen-forming fungi. J. Phycol. 47:282-290.

53. Ogle, H. 1997. Other biotic causes of plant diseases. Pages 143-145 in: Plant Pathogens and Plant Diseases. J. F. Brown and H. J. Ogle, eds. Australas. Plant Path. Soc.,Australia.

54. Ong, B. L., Lim, M., and Wee, Y. C. 1992. Effects of desiccation and illumination on photosynthesis and pigmentation of an edaphic population of Trentepohlia odorata (Chlorophyta). J. Phycol. 28:768-772.

55. Pautasso, M., Döring, T. F., Garbelotto, M., Pellis, L., and Jeger, M. J. 2012.Impacts of climate change on plant diseases—opinions and trends. Eur. J. Plant Pathol. 133:295-313.

56. Ponmurugan, P., Saravanan, D., and Ramya, M. 2010. Culture and biochemical analysis of a tea algal pathogen, Cephaleuros parasiticus. J. Phycol. 46:1017-1023.

57. Printz, H. 1939. Vorarbeiten zu einer Monographie der Trentepohliaceen. Nytt Magasin for Naturvidenskapene 80:137-210.

58. Ramya, M., Ponmurugan, P., and Saravanan, D. 2013. Management of Cephaleuros parasiticaus [sic] Karst (Trentepohliales: Trentepohliaceae), an algal pathogen of tea plant, Camellia sinsensis [sic] (L) (O. Kuntze). Crop Prot. 44:66-74.

59. Reynolds, D. R., and Dunn, P. H. 1982. Foliicolous fungi from Brazil: the new genusVeralucia. Mycologia 74:854-857

60. Reynolds, D. R., and Dunn, P. H. 1984. A fungus-like alga. Mycologia 76:719-721.

61. Rindi, F., and Guiry, M. D. 2002. Diversity, life history and ecology of Trentepohlia andPrintzina (Trentepohliales, Chlorophyta) in urban habitats in Western Ireland. J. Phycol. 38:39-54.

62. Rindi, F., Guiry, M. D., and López-Bautista, J. M. 2006. New records of Trentepohliales (Ulvophyceae, Chlorophyta) from Africa. Nova Hedwigia 83:431-449.

63. Rindi, F., Lam, D. W., and López-Bautista, J. M. 2009. Phylogenetic relationships and species circumscription in Trentepohlia and Printzina (Trentepohliales, Chlorophyta). Mol. Phylogenet. Evol. 52:329-339.

64. Salleh, A., and Kamsare, S. 1994. Studies on Cephaleuros virescens Kunze, a parasitic alga from Malaysia. Pages 274-278 in: Algal Biotecnol. in the Asia-Pacific Region. S. M. Prang, Y. K. Lee, M. Borowotzka, and B. A. Whitton, eds. University of Malaysia,Kuala Lumpur.

65. Sarma, P. 1986. The freshwater Chaetophorales of New Zealand. Nova Hedwigia Beih. 58:1-169.

66. Schoch, C. L., Sung, G. H., López-Giráldez, F., Townsend, J. P., Miadlikowska, J.,Hofstetter, V., and Shoemaker, R. A. 2009. The Ascomycota tree of life: a phylum-wide phylogeny clarifies the origin and evolution of fundamental reproductive and ecological traits. Syst. Biol. 58:224-239.

67. Schubert, T. S. 1981. Strigula Fries, the plant parasitic lichen. Plant Pathology Circular No. 227, Fla. Dept. Agric. Consumer Serv., Div. Plant Ind., Tallahassee, FL.

68. Suto, Y., Ganesan, E. K., and West, J. A. 2014. Comparative observations onCephaleuros parasiticus and C. virescens (Trentepohliaceae, Chlorophyta) from India. Algae 29:121-126

69. Suto, Y., and Ogawa, S. 1988. Diseases in Camellia japonica and C. sasanqua (1). Page 108 in: Diseases of Ornamental and Forest Trees. T. Kobayashi, ed. Yokendo, Tokyo (in Japanese).

70. Suto, Y., and Ohtani, S. 2009. Morphology and taxonomy of five Cephaleuros species (Trentepohliaceae, Chlorophyta) from Japan, including three new species. Phycologia 48:213-236.

71. Suto, Y., and Ohtani, S. 2011a. The Strigula smaragdula complex (lichenized Ascomycota, Strigulaceae) on living leaves of woody plants from Shimane-ken, western Japan. Lichenology 10:1-13.

72. Suto, Y., and Ohtani, S. 2011b. Morphological features and chromosome numbers in cultures of five Cephaleuros species (Trentepohliaceae, Chlorophyta) from Japan. Phycol. Res. 59:42-51.

73. Suto, Y., and Ohtani, S. 2011c. Inoculations of the aerial algae of five Cephaleurosspecies to living leaves of various woody plants. Jpn. J. Phycol. Sorui 59:131-138 (in Japanese with English summary).

74. Suto, Y., and Ohtani, S. 2013. Seasonal development of five Cephaleuros species (Trentepohliaceae, Chlorophyta) on the leaves of woody plants and the behaviors of their gametes and zoospores. Phycol. Res. 61:105-115.

75. Thompson, R. H., and Wujek, D. 1997. Trentepohliales: Cephaleuros, Phycopeltis, andStomatochroon: Morphology, Taxonomy, and Ecology. Science Publishers Inc., Enfield, NH.

76. Timpano, P., and Pearlmutter, N. L. 1983. Algal invasion of angiosperm mesophyll. Scan. Electron Microsc. 86:1511-1517.

77. Van den Hoek, C., Mann, D. G., and Jahns, H. M. 1995. Algae: An Introduction to Phycology. Cambridge University Press, Cambridge, UK.

78. Ward, H. M. 1884. Structure, development, and life-history of a tropical epiphyllous lichen (Strigula complanta Fée).Trans. Linn. Soc. London, Ser. 2. Botany 2:87-119.
Wellman, F. L. 1965. Pathogenicity of Cephaleuros virescens in the Neotropics. Phytopathology 55:283-286.

79. Wolf, F. A. 1930. A parasitic alga, Cephaleuros virescens Kunze, on citrus and certain other plants. J. Elisha Mitch. Sci. Soc. 45:187-205.

80. Zechman, F. W., Theriot, E. C., Zimmer, E. A., and Chapman, R. L. 1990.Phylogeny of the Ulvophyceae (Chlorophyta): cladistic analysis of nuclear-encoded rDNA sequence data. J. Phycol. 26:700-710.

81. Zhu, H., Zhao, Z.-J., Xia, S., Hu, Z.-Y., and Liu, G.-X. 2014. Morphology and phylogenetic position of Stomatochroon reniformis var. chinensis var. nov. (Trentepohliales, Ulvophyceae), a rare endobiotic alga from China. Phycologia 53:493-501.

Chapter 11

Development of Molecular Assays for Detection of *Stenocarpella maydis* and*Stenocarpella macrospora* in Corn

M. P. Romero and K. A. Wise

Department of Botany and Plant Pathology, Purdue University, West Lafayette, IN 47907

ABSTRACT

The causal agents of Diplodia ear rot are two species of the *Stenocarpella* genus, *S. macrospora* and *S. maydis*. In addition to ears, both pathogens can infect leaves and stalks, and both are present in most corn-production regions around the world. It is difficult to visually distinguish between the two pathogens based on plant symptoms and fungal signs. To facilitate accurate and rapid pathogen identification, polymerase chain reaction (PCR) assays were developed for identification of each species. Species-specific primers of 18 to 20 nucleotides in length were designed, targeting a portion of the internal transcribed spacer (ITS) region of the fungal genome for conventional and real-time PCR assays. The conventional PCR method successfully amplified a single 1.7-kb and 800-bp fragment for each *S. maydis* and *S. macrospora*isolate, respectively. The real-time method was performed using SYBR green dye, and detection of each specific target pathogen was successfully obtained. In total, 82 *S. maydis* and 15 *S. macrospora* isolates were tested to evaluate the reproducibility of these primers. Both methods provide a rapid and specific tool for the detection ofStenocarpella spp.

Stenocarpella macrospora (Earle) B. Sutton and *S. maydis* (Berk.) B. Sutton, the only two species included in the *Stenocarpella* genus (9,12,47,48), have been identified as causal agents of Diplodia ear rot, Diplodia leaf streak, and Diplodia stalk rot (53). Both species are able to infect ears, stalks, and leaves (47,48) and, under specific conditions, these fungi can produce mycotoxins (10,45,55). *S. maydis* is recognized as one of the most persistent and important ear rot pathogens in the world (41). Typically, *S. macrospora* has been more significant in tropical and subtropical areas (15,20,24); however, the disease is commonly present in the United States, and increases in disease incidence have been observed across the Midwest in recent years (3,52).

Visually distinguishing between symptoms caused by these two species is challenging (6).*S. macrospora* produces long elliptical gray-green lesions with yellowish margins on leaves. Concentric zones are evident in the lesion around the point of infection origin. Pycnidia are typically observed in the older part of the lesion (19,24,53). Foliar lesions caused by *S. maydis* are oval and smaller than those caused by *S. macrospora* (19) and pycnidia production is less abundant across the lesion. Both diseases may be confused with other common corn foliar diseases, including northern corn leaf blight (caused by*Exserohilum turcicum*; 1) and Goss's bacterial leaf blight (*Clavibacter michiganensis* subsp.*nebraskensis*; 56). When ears are infected, both species can colonize the ear with white, cottony mycelia. Black pycnidia can be observed on the husk or on the kernel surface (6,13,24). Crop rotation and tillage are currently the only reliable methods for disease management. Hybrids with complete resistance to these diseases are not known, although some hybrids may be less susceptible than others.

In 2012, Diplodia ear rot, leaf streak, and stalk rot resulted in a combined loss of 0.97 million metric tons in the United States and Ontario, Canada (31). In addition to yield loss, in some subtropical and tropical corn production areas, cases of diplodiosis, a neuromycotoxicosis of livestock, have been reported and associated with *S. maydis* (17,23,28,35,37). Diplodiatoxin and diplonine have also been associated with diplodiosis but these mycotoxins do not always cause the neurological signs associated with diplodiosis symptoms in livestock animals and, therefore, it is believed that the main causal agents responsible for this mycotoxicosis are still unknown (45,46,55). *S. macrospora* produces diplosporin and diplodiol, two mycotoxins that affect small animals (10,18). Field outbreaks were reported only in South Africa (16,28) until 1964, when Darvall reported a possible outbreak in Australia. Recently, Argentina (32) and Brazil (38) reported cases of animal poisoning by diplodiosis. Although there have been no reports of diplodiosis in the United States (22,51), corn grain is tested for the presence of these pathogens when exported to other countries. *S. maydis* is considered a regulated nonquarantine pest (RNQP) in Brazil, due to the negative impact it has on corn production (2,44).

Currently, *S. maydis* is considered a major pathogen throughout the corn-producing regions in the United States and, although *S. macrospora* is currently considered a minor pathogen in the United States, Latterell et al. (19) concluded that this pathogen could compromise corn production in humid areas of the United States. The distribution and prevalence of *S. macrospora* in the United States is not well known. Previous studies have focused on determining morphological differences between these two pathogens. There are clear differences in conidial dimensions and number of septa between the two species. *S. maydis* conidia are 15 to 34 by 5 to 8 µm, with 0 to 2 septae. *S. macrospora* conidia are almost triple the size of *S. maydis* conidia, being 55 to 106 by 6 to 14 µm and with 0 to 3 septae (12,24,47,48). There are also color variances between young and old cultures of each species when grown on potato dextrose agar (PDA) (8,25,36). Initially, both species have a white, cottony appearance on the media; however, after 2 weeks, *S. maydis*cultures turn grayish to dark brown. Cultures of *S. macrospora* maintain the original white appearance. Another distinguishable characteristic is that *S. maydis* reportedly produces higher numbers of pycnidia in culture (30). Pycnidia production of *S. macrospora* in culture increases when mycelium is removed from surface of culture. However, *S. macrospora*requires more time to develop pycnidia and, frequently, pycnidia are embedded in the media or at the edge of the plate (15,19,29), making it difficult to obtain conidia for identification (2).

Previous studies described the use of the polymerase chain reaction (PCR) method to identify *Stenocarpella* spp. (33,57). Primers designed for these assays were based on the internal transcribed spacer (ITS) region for *Stenocarpella* spp., and these assays were able to differentiate *Stenocarpella* spp. from other fungal species but were not specific enough to differentiate between the two *Stenocarpella* spp. (2).

According to Lievens et al. (21), the absence of an effective and accurate method to detect and identify pathogens is one of the major limitations of plant disease management. Although *Stenocarpella* spp. can be distinguished based on a combination of morphological and cultural factors, these techniques are time consuming and are not conducive to the accurate and rapid diagnosis required for regulatory agencies that monitor corn seed pathogens to prevent pathogen establishment in new areas and to monitor grain for the presence of pathogens that produce mycotoxins. *S. maydis* is considered an RNQP in Brazil, due to the negative impact it has on corn production, and corn is routinely screened for presence of the pathogen (2, 44). In fact, Casa et al. (6) reported that seed infected with *Stenocarpella* spp. was responsible for introducing the pathogen into areas in Brazil where it was not previously established. Currently, molecular assays for detecting and distinguishing *Stenocarpella* spp. are lacking. The objective of this study was to design and validate conventional and real-time PCR assays for *Stenocarpella* spp.

1. MATERIALS AND METHODS

1.1 Fungal isolates and growth conditions

In total, 82 *S. maydis* isolates and 15 *S. macrospora* isolates collected in the United States were included in this study (Table 1). Other pathogenic species that commonly infect corn or are closely related to the *Stenocarpella* genus were included to assess assay validity (Table 1). *Stenocarpella* spp. isolates were obtained from corn kernels and corn leaves collected between 2010 and 2014 from fields in Illinois, Indiana, Missouri, New York, Tennessee, and Wisconsin. Isolates were obtained from plant tissue by disinfecting leaves with 5% commercial bleach (a.i. 5.05% sodium hypochlorite) and Tween 20 (Bio-Rad Laboratories, Hercules, CA) for 10 s. Tissue was washed twice with water, dried for 4 h, and transferred to PDA. Cultures were stored at 28°C under 12-h light and dark cycles until pycnidia were visible (approximately 7 days). A single pycnidium was hand picked with a sterile needle and placed into a 1.5-ml Flex-Eppendorf tube (Eppendorf AG, Hamburg, Germany) with 150 µl of double distilled H_2O. After agitation, four drops of the suspension were transferred with a sterilized pipette to water agar, containing Bacto agar (Becton, Dickinson and Company, Sparks, MD) at 20 g/liter. Plates were incubated at room temperature for 24 h under dark conditions. Single germinated spores were picked and transferred to natural oatmeal agar (NOA) (40).

Plates were stored at 28°C under 12-h light and dark cycles until the plate was fully colonized (approximately 7 days). Once a single-spore culture was obtained, a 5-mm plug was transferred to a new NOA plate already containing sterilized filter paper (8 by 8 cm). Plates were placed again at 28°C under 12-h light and dark cycles until the filter paper was fully colonized and pycnidia production was visible (approximately 15 days). Filter paper was removed from the plate with a sterile forceps and transferred to new, empty petri dish. The filter paper was completely dried, placed inside an envelope covered with foil paper, and stored at −20°C until further use.

Table 1. Isolate code, year, and collection site of *Stenocarpella* spp. and other pathogenic species used in this study

Isolate code	Pathogen	Year	County, state
1-1-1	*Stenocarpella maydis*	2010	Tippecanoe, IN
1-1-3	*S. maydis*	2010	Tippecanoe, IN
1-2-4	*S. maydis*	2010	Tippecanoe, IN
1-3-3	*S. maydis*	2010	Tippecanoe, IN
1-6-2	*S. maydis*	2010	Tippecanoe, IN
1-11-2	*S. maydis*	2010	Tippecanoe, IN
1-11-4	*S. maydis*	2010	Tippecanoe, IN
1-13-2	*S. maydis*	2010	Tippecanoe, IN
1-14-4	*S. maydis*	2010	Tippecanoe, IN
1-18-1	*S. maydis*	2010	Tippecanoe, IN
1-19-1	*S. maydis*	2010	Tippecanoe, IN
1-20-4	*S. maydis*	2010	Tippecanoe, IN
2-2-1	*S. maydis*	2010	Tippecanoe, IN
2-3-1	*S. maydis*	2010	Tippecanoe, IN
2-4-1	*S. maydis*	2010	Tippecanoe, IN
2-6-2	*S. maydis*	2010	Tippecanoe, IN
2-7-4	*S. maydis*	2010	Tippecanoe, IN
2-8-1	*S. maydis*	2010	Tippecanoe, IN
2-13-1	*S. maydis*	2010	Tippecanoe, IN
2-14-1	*S. maydis*	2010	Tippecanoe, IN
2-16-1	*S. maydis*	2010	Tippecanoe, IN
2-20-1	*S. maydis*	2010	Tippecanoe, IN
3-1-1	*S. maydis*	2010	Tippecanoe, IN
3-2-2	*S. maydis*	2010	Tippecanoe, IN
3-3-1	*S. maydis*	2010	Tippecanoe, IN
3-3-2	*S. maydis*	2010	Tippecanoe, IN
3-4-4	*S. maydis*	2010	Tippecanoe, IN

3-5-2	*S. maydis*	2010	Tippecanoe, IN
3-6-4	*S. maydis*	2010	Tippecanoe, IN
3-8-2	*S. maydis*	2010	Tippecanoe, IN
3-9-1	*S. maydis*	2010	Tippecanoe, IN
3-9-3	*S. maydis*	2010	Tippecanoe, IN
3-9-4	*S. maydis*	2010	Tippecanoe, IN
3-10-1	*S. maydis*	2010	Tippecanoe, IN
3-10-2	*S. maydis*	2010	Tippecanoe, IN
3-11-1	*S. maydis*	2010	Tippecanoe, IN
3-11-2	*S. maydis*	2010	Tippecanoe, IN
3-12-3	*S. maydis*	2010	Tippecanoe, IN
3-12-4	*S. maydis*	2010	Tippecanoe, IN
1b.1	*S. maydis*	2011	Tippecanoe, IN
1b.2	*S. maydis*	2011	Tippecanoe, IN
2b.2	*S. maydis*	2011	Tippecanoe, IN
3a.1	*S. maydis*	2011	Tippecanoe, IN
4b.1	*S. maydis*	2011	Tippecanoe, IN
7a.1	*S. maydis*	2011	Tippecanoe, IN
8b.1	*S. maydis*	2011	Tippecanoe, IN
8b.2	*S. maydis*	2011	Tippecanoe, IN
9b.1	*S. maydis*	2011	Tippecanoe, IN
10a.1	*S. maydis*	2011	Tippecanoe, IN
10a.2	*S. maydis*	2011	Tippecanoe, IN
4a.1.1	*S. maydis*	2011	Indiana
4a.1.2	*S. maydis*	2011	Indiana
3b.1	*S. maydis*	2011	Indiana
1.2	*S. maydis*	2012	Tippecanoe, IN
1.4	*S. maydis*	2012	Tippecanoe, IN
2	*S. maydis*	2012	Tippecanoe, IN
2.1	*S. maydis*	2012	Tippecanoe, IN
3.1	*S. maydis*	2012	Tippecanoe, IN
3.2	*S. maydis*	2012	Tippecanoe, IN
3.3	*S. maydis*	2012	Tippecanoe, IN
3.4	*S. maydis*	2012	Tippecanoe, IN
4.1	*S. maydis*	2012	Tippecanoe, IN
5.1	*S. maydis*	2012	Tippecanoe, IN
5.2	*S. maydis*	2012	Tippecanoe, IN
6	*S. maydis*	2012	Tippecanoe, IN
7.1	*S. maydis*	2012	Tippecanoe, IN
8	*S. maydis*	2012	Tippecanoe, IN
9	*S. maydis*	2012	Tippecanoe, IN
5	*S. maydis*	2013	Tippecanoe, IN
8	*S. maydis*	2013	Tippecanoe, IN
10	*S. maydis*	2013	Knox, IN
13	*S. maydis*	2013	Knox, IN
15.1	*S. maydis*	2013	Howard, MO
15.2	*S. maydis*	2013	Howard, MO
44	*S. maydis*	2013	Cayugo, NY
45	*S. maydis*	2013	Tippecanoe, IN
1	*S. maydis*	2014	Gibson, TN
2	*S. maydis*	2014	Gibson, TN
3	*S. maydis*	2014	Gibson, TN

5	*S. maydis*	2014	Gibson, TN
6	*S. maydis*	2014	Gibson, TN
7	*S. maydis*	2014	Gibson, TN
1a.1.1	*S. macrospora*	2011	Indiana
1a.1.2	*S. macrospora*	2011	Indiana
P11-2.1a.1	*S. macrospora*	2011	Kosciusko, IN
P11-2.1a.2	*S. macrospora*	2011	Kosciusko, IN
4a.2	*S. macrospora*	2011	Indiana
4c.1.1	*S. macrospora*	2011	Indiana
10a.2	*S. macrospora*	2011	Indiana
10.1b	*S. macrospora*	2011	Indiana
1a	*S. macrospora*	2012	Tippecanoe, IN
1b	*S. macrospora*	2012	Tippecanoe, IN
4b	*S. macrospora*	2012	Tippecanoe, IN
1	*S. macrospora*	2013	Indiana
2.1a	*S. macrospora*	2013	Jennings, IN
2.2	*S. macrospora*	2013	Jennings, IN
4	*S. macrospora*	2014	Gibson, TN
ET	*Exserohilum turcicum*	2012	Tippecanoe, IN
Fv	*Fusarium verticillioides*	2012	Indiana
Fg	*Fusarium graminearum*	2012	Indiana
Af	*Aspergillus flavus*		Urbana, IL
Cmn	*Clavibacter michiganensis* subsp. *Nebraskensis*	2008	Pulaski, IN
Mp	*Macrophomina phaseolina*	2011	Grant, WI
Cg	*Colletotrichum graminicola*	2012	Illinois
Ps	*Phoma* spp.	2012	Hamilton, IN
Dp	*Diplodia pinea*	2012	Tippecanoe, IN

1.2 DNA extraction

Pure cultures from storage were subcultured on fresh NOA plates. Plates were incubated at 28°C under 12-h light and dark cycles for 7 days, after which approximately 40 mg of fresh mycelium was collected by scraping from the colony surface. DNA was extracted from each isolate by a cetyltrimethyl ammonium bromide (CTAB) protocol derived from Saghai-Maroof et al. (42) and stored in nuclease-free sterile water at −20°C. DNA also was extracted from corn kernels and stalks infected by *S. maydis* in 2014 and from kernels and leaves infected by *S. macrospora* in 2014. DNA extraction from corn plant tissue also followed the CTAB protocol. DNA concentration was measured with a Nanodrop 1000 spectrophotometer (Thermo Fisher Scientific, Wilmington, DE). For conventional and real-time PCR assays, DNA concentrations were adjusted to 10 ng/µl.

1.3 ITS PCR and sequencing

The ITS regions of *Stenocarpella* spp. were amplified with ITS1F (CTTGGTCATTTACAGGAAGTAA; 14) and ITS4 (TCCTCCGCTTATTGATATGC; 54) primers. PCR amplifications were carried out in a 25-µl reaction volume containing 1 µl of template DNA (various DNA concentrations), 0.20 mM each deoxynucleoside triphosphate (Promega Corp., Madison WI), 4 U of Taq DNA polymerase (New England BioLabs Inc., Ipswich, MA), 1× PCR buffer (New England BioLabs Inc.), 0.60 µM each forward and reverse primer, and 16 µl of nuclease-free sterile water. PCR was conducted with a AB 2720 Thermal Cycler (Applied Biosystems, Foster City, CA) with the following program: an initial heating cycle of 95°C for 2 min; 30 cycles of denaturation at 95°C for 15 s, annealing at 55°C for 30 s, and extension at 72°C for 1.5 min; and a 10-min final extension at 72°C. Each total product volume (10 µl) was loaded on a 1% (wt/vol) agarose gel in 150 1× Tris-acetate-EDTA buffer and stained with ethidium bromide. The gels were run at 50 V for 40 min and visualized using a UV-transilluminator FBTI-88 (Thermo Fisher Scientific Inc., Waltham, MA). PCR products were extracted from gels, purified, and sent to the Purdue Genomics Facility for sequencing. DNA sequences were BLAST analyzed to a database constructed from a partial genome assembly. The ITS region was located on a 7.8-kb contig (GenBank accession number KP164561). The sequence of this contig was used to identify two specific sets of primers for *S. maydis* for both conventional and real-time PCR assays. The sequence of the amplified fragments from S. *macrospora* was aligned with the *S. maydis*sequence, and a single set of primers for this species was designed to use in both conventional and real-time PCR assays. Both species-specific sets of primers (Table 2) were synthesized by Integrated DNA Technologies (Coralville, IA).

1.4 Conventional PCR for the diagnosis of *Stenocarpella* spp

To determine the optimal annealing temperature of each primer set, temperature gradient PCR was carried out using a Mastercycler pro S (Eppendorf AG) with an annealing temperature range of 52 to 63°C for *S. macrospora* and 54 to 65°C for *S. maydis*. The optimal annealing temperatures for *S. macrospora* and *S. maydis* were 57 and 60°C, respectively. One negative control, containing nuclease-free sterile water instead of DNA, was included with each species assay.

Table 2. Primer sequence used to identify isolates of *Stenocarpella*spp.

Primer	Sequence 5′–3′	Annealing temperature (°C)	Product size (bp)	Product melt curve peak (°C)
Smay.F	CCTGCTATGCATAGGTCG	52.6	...	...
Smay.R	CAC CAG GCC GTT AAG CCT TA	57.4	1,700	...
RT.Smay.F	GTT TCC ATG ACC TGC TCA CG	56.3	...	...
RT.Smay.R	TGT TGC TCG GTT TCA GGC TTG	58.4	500	88
Smac.F	GGG CAA ATT TTC TCG GAG G	54	...	...
Smac.R	GCA GCT ATT CAG CGT TCA TC	54.1	800	...
RT.Smac.F	GGG CAA ATT TTC TCG GAG G	54	...	...
RT.Smac.R	GCA GCT ATT CAG CGT TCA TC	54.1	800	89

For *S. maydis*, PCR were conducted in a 20-µl reaction volume containing template DNA at 10 ng/µl, 0.25 mM each deoxynucleoside triphosphate (Promega Corp.), Taq DNA polymerase (New England BioLabs Inc.) at 4 U/µl, 1× PCR buffer (New England BioLabs Inc.), 0.75 µM each forward and reverse primer (Table 2), and 11 µl of nuclease-free sterile water (Thermo Fisher Scientific Inc., Waltham, MA). Fragments were amplified using the Smay.F/Smay.R primers (Table 2) under the following program: an initial heating cycle of 95°C for 2 min; 35 cycles of denaturation at 95°C for 30 s, annealing at 60°C for 30 s, and extension at 72°C for 1 min; and a 10-min final extension at 72°C.

For *S. macrospora,* PCR were conducted in a 20-µl reaction volume containing template DNA at 10 ng/µl, 0.25 mM each deoxynucleoside triphosphate (Promega Corp.), Taq DNA polymerase (New England BioLabs Inc.) at 4 U/µl, 1× PCR buffer (New England BioLabs Inc.), 0.25 µM forward primer (Table 2), 0.75 µM reverse primer (Table 2), and 12 µl of nuclease-free sterile water (Thermo Fisher Scientific Inc., Waltham, MA). Fragments were amplified using the Smac.F/Smac.R primers (Table 2) under the following program: an initial heating cycle of 95°C for 2 min; 30 cycles of denaturation at 95°C for 30 s, annealing at 57°C for 30 s, and extension at 72°C for 1 min; and a 10-min final extension at 72°C.

Following PCR, amplified products were separated via gel electrophoresis and visualized as described above. Each reaction was performed three times. All amplified PCR products were resequenced for verification.

1.5 Real-time PCR for the diagnosis of *Stenocarpella* spp

PCR assays were performed using a LightCycler 480 Instrument (Roche Diagnostics Corporation, Hague Road, IN). For *S. maydis*, real-time PCR was conducted in a 20-µl volume reaction mix which included template DNA at 10 ng/µl, 1× iTaq Universal SYBR Green Supermix (Bio-Rad Laboratories), 0.10 µM forward primer, 0.60 µM reverse primer, and 8.6 µl of nuclease-free sterile water (Thermo Fisher Scientific Inc., Waltham, MA). Isolates tested using the RT.Smay.F/RT.Smay.R primers were run as follows: 95°C for 2 min, followed by 35 cycles of 95°C for 30 s, 60°C for 30 s, and 72°C for 1 min. The thermal profile for melting curve analysis consisted of a denaturation for 5 s at 95°C, lowered to 55°C for 1 min, then increased to 97°C with continuous fluorescence readings.

In total, 15 isolates of *S. macrospora* were amplified in a reaction volume of 20 µl which included template DNA at 10 ng/µl, 1× iTaq Universal SYBR Green Supermix (Bio-Rad Laboratories), 0.25 µM forward primer, 0.75 µM reverse primer, and 7 µl of nuclease-free sterile water (Thermo Fisher Scientific Inc., Waltham, MA). These 15 isolates were tested with the same primers (RT.Smac.F/RT.Smac.R) and thermal cycling settings as those for the conventional PCR. The thermal profile for melting curve analysis was the same as forS. *maydis*.

Water control (nuclease-free sterile water) was included in each run for both assays. To evaluate specificity of each assay, the heterologous *Stenocarpella* spp., six fungal pathogens of corn, one bacterial pathogen of corn, and two pathogenic fungi not known to attack corn were also included in each run (Table 1). Each reaction was performed three times.

1.6 Detection limit for conventional and real-time PCR

The detection limit for each assay was determined by serial 10-fold dilution of genomic DNA from pure fungal cultures and from corn seed. Dilutions ranged from 10 to 10×10^{-6}ng/µl for *S. maydis* and 10 to 10×10^{-3} ng/µl for *S. macrospora*. The PCR conditions for both assays described previously were used. Amplification was confirmed on a 1% (wt/vol) agarose gel and by melting curve profile for conventional and real-time PCR, respectively.

2. RESULTS

Fragments of the ITS region were successfully isolated from the two *Stenocarpella* spp. using primers ITS1F and ITS4 (gel not shown). The primer set Smay.F/Smay.R, specific for *S. maydis*, generated a single and specific PCR product of 1.7 kb (Fig. 1A). This PCR product was observed in all 82 *S. maydis* isolates tested and in none of the *S. macrospora*isolated tested. No cross-amplification product was obtained when DNA from other fungal or bacterial pathogens was used (Fig. 1B). When DNA extracted directly from infected corn kernels and stalks (not pure culture) was tested using the conventional PCR set of primers, a specific PCR product of 1.7 kb was obtained (gel not shown).

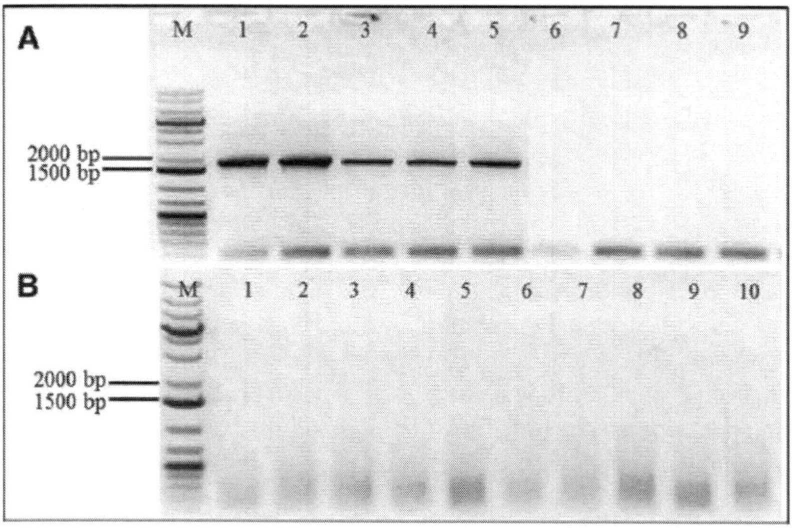

Figure 1. Agarose gel electrophoresis of polymerase chain reaction products. **A,** Products of *Stenocarpella maydis* obtained with primers Smay.F/Smay.R. Lanes 1 to 5: isolates 3-10-1, 3a.1, 4a.1.2, 7.1, and 10a.2, respectively; lane 6 to 8: *S. macrospora* isolates 1a.1.1, 1a, and 4b, respectively; lane 9: control (nuclease-free sterile water); lane M: molecular size marker (Life Technologies, Grand Island, NY). **B,** Products of other fungal and bacteria species obtained with primers Smay.F/SmayR. Lanes 1 to 9: *Phoma*spp., *Diplodia pinea, Fusarium graminearum, Aspergillus flavus, Clavibacter michiganensis* subsp. *nebraskensis, Exserohilum turcicum, Colletotrichum graminicola, Macrophomina phaseolina*, and *Fusarium verticillioides*, respectively; lane 10: control (nuclease-free sterile water; lane M: molecular size marker (Life Technologies).

The *S. macrospora*-specific primers Smac.F/Smac.R generated a single PCR product of 800 bp (Fig. 2A). This PCR product was observed only in the 15 *S. macrospora* isolates tested, and DNA extracted from other fungal or bacterial isolates did not generate a PCR product (Fig. 2B). When DNA extracted directly from infected corn kernels and leaves (not pure culture) was tested using the conventional PCR set of primers, a specific PCR product of 800 bp was observed on the corn kernels DNA; however, for the DNA extracted from corn leaves, a faint band was obtained (gel not shown).

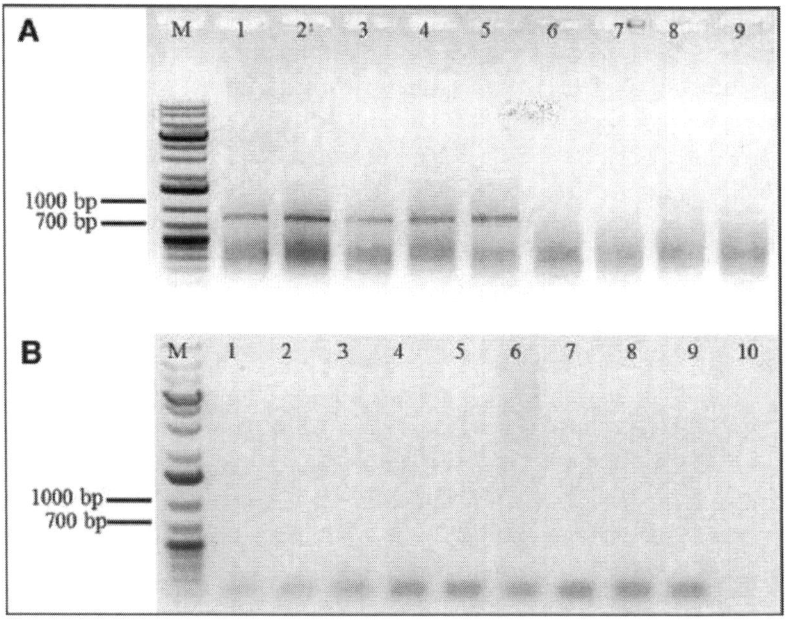

Figure 2. Agarose gel electrophoresis of polymerase chain reaction products. **A,** Products of *Stenocarpella macrospora* obtained with primers Smac.F/Smac.R. Lanes 1 to 5: isolates 1b, 4a.2, 4b, 1a.1.1, and 1, respectively; lane 6 to 8: *S. maydis* isolates 2-2-1, 7.1, and 3a.1, respectively; lane 9: control (nuclease-free sterile water; lane M: molecular size marker (Life Technologies). **B,** Products of other fungal and bacteria species obtained with primers Smac.F/Smac.R. Lanes 1 to 9: *Phoma* spp., *Diplodia pinea, Fusarium graminearum, Aspergillus flavus, Clavibacter michiganensis* subsp. *nebraskensis, Exserohilum turcicum, Colletotrichum graminicola, Macrophomina phaseolina*, and *Fusarium verticillioides*, respectively; lane 10: control (nuclease-free sterile water; lane M: molecular size marker (Life Technologies).

Primers designed for the real-time assay were tested first in a conventional PCR assay. Two PCR products of 500 and 800 bp were generated by the primer sets RT.Smay.F/RT.Smay.R and RT.Smac.F/RT.Smac.R, respectively, for each target-species (gel not shown). No cross-amplification product was obtained when DNA from other fungal or bacterial isolates was tested.

The real-time PCR assay for *S. maydis* specifically amplified DNA from all *S. maydis*isolates tested (Fig. 3A). Positive *S. maydis* isolates had a range of melting peaks of 88.54 to 88.94°C (Fig. 3B and C) and cycle threshold (C_t) values of 23.18 to 26.16 (Fig. 3C). DNA extracted from corn kernels and stalks infected with *S. maydis* were tested with RT.Smay.F/RT.Smay.R primers, melting peaks were 88.74 and 88.94°C, respectively (Fig. 3B and C), and C_t values were 28.55 and 30.00, respectively (Fig. 3C). The *S. maydis*-specific real-time PCR assay did not amplify DNA of any of the *S. macrospora* isolates. Negative control samples showed a different range of melting peaks (64 to 78.01°C; Fig. 3B and C) compared with the *S. maydis* isolates. C_t values from most of the negative controls were ≥30; however, DNA of the nontarget species *Aspergillus flavus* and *E. turcicum,* reported C_t values of 26.08 and 27.21, respectively. The specific primers RT.Smac.F/RT.Smac.R produced an amplicon of the expected size for all *S. macrospora*isolates (Fig. 4A). *S. macrospora* isolates had a melting peak range of 89.39 to 89.80°C (Fig. 4B and C), and C_t values of 18.31 to 21.69 (Fig. 4C). DNA extracted from infected corn kernels and leaves with *S. macrospora* were tested with RT.Smac.F/RT.Smac.R primers, melting peaks were 89.18 and 88.25°C, respectively, and C_t values were 22.31 and 23.89, respectively (data not shown). The *S. macrospora*-specific assay did not amplify DNA from any *S. maydis* isolates or negative controls included in the assay (melting peak range of 76.74 to 83.28°C and C_t values = 25.00; Fig. 4B and C).

Detection limits of *S. maydis* in conventional and real-time PCR were 0.01 ng/µl with Smay.F/SmayR primers (Fig. 5A) and 0.001 ng/µl with RT.Smay.F/RT.Smay.R primers (Fig. 5A). The detection limits of *S. macrospora* in conventional and real-time assays were 0.1 ng/µl with Smac.F/Smac.R primers (Fig. 5B), and 1 ng/µl with RT.Smac.F/RT.Smac primers (Fig. 5B). Detection limits of DNA extracted from infected corn seed were 10 ng/µl for *S. maydis* and *S. macrospora* in conventional PCR, and 1 ng/µl for *S. maydis* and *S. macrospora* in real-time PCR (data not shown).

3. DISCUSSION

This is the first study to successfully develop multiple PCR assays with the specificity to distinguish *S. maydis* from *S. macrospora*. These pathogens are difficult to distinguish based on pathogen signs and disease symptoms, and previous diagnostic assays have been time consuming or nonspecific. Both conventional and real-time assays were able to distinguish between *S. maydis* and *S. macrospora* isolates collected from several states in the United States Corn Belt.

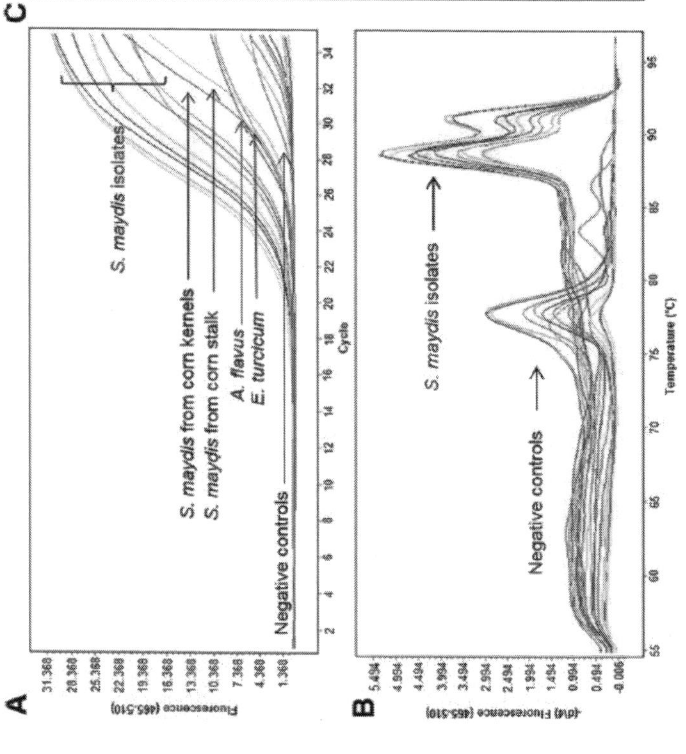

C

Isolate[a]	Ct[b]	Temperature
3b.1	25.98±0.15	88.72
3a.1	24.82±0.16	88.89
4b.1	26.16±0.04	88.90
10a.2	23.18±0.26	88.70
3.1	24.37±0.18	88.54
9	24.71±0.03	88.91
5.2	24.10±0.19	88.66
4.1	23.31±0.30	88.64
Corn kernel	28.55±0.41	88.74
Corn stalk	30.00	88.84
A. flavus	27.21±0.20	64.26
C. graminicola	30.00	78.80
E. turcicum	28.08±0.02	77.73
F. graminearum	30.00	77.64
F. verticillioides	30.00	73.85
Cmn[c]	30.00	78.01
M. phaseolina	30.00	77.43
O. pinnea	30.00	77.61
Phoma spp.	30.00	77.68
S. macrospora	30.00	77.45
Water	30.00	77.65

Figure 3. Melting curve analysis for *Stenocarpella maydis* using real-time polymerase chain reaction (PCR) assay. **A,** Amplification curves of *S. maydis* isolates and controls obtained with primers RT.Smay.F/RT.Smay.R. **B,** Melting curve analysis of the same samples shows the presence of the specific PCR product. **C,** Comparison chart with cycle threshold (C_t) values and melting temperatures. Superscript a: DNA extracted for other fungi isolated that are not the target of this assay and water are identified as negative controls. Superscript b: C_t values ± standard deviation. Superscript c: Cmn = *Clavibacter michiganensis* subsp. *nebraskensis*.

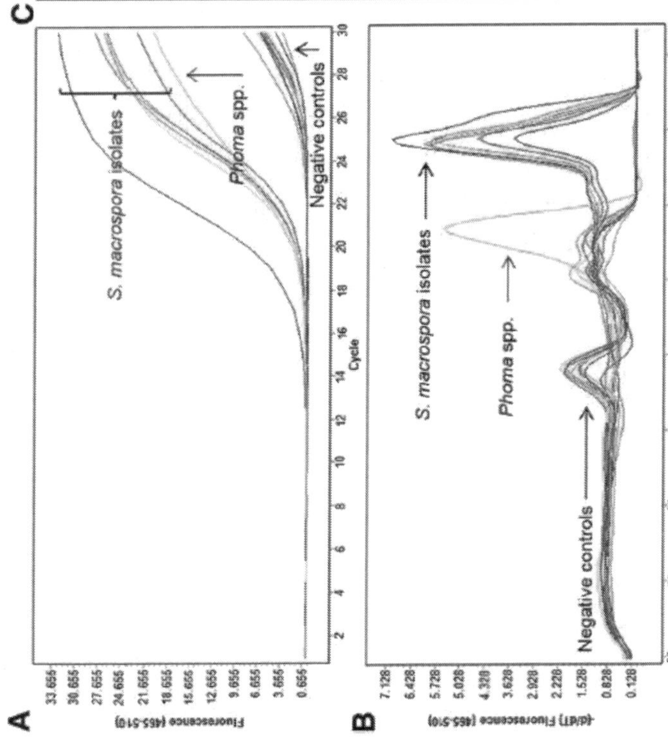

C

Isolate[a]	Ct[b]	Temperature
P11-2.1a.2	21.24±0.03	89.39
2.1a	21.69±0.41	89.60
1a.1.1	21.11±0.09	89.40
4c.1.1	21.20±0.49	89.71
10.1b	21.15±0.51	89.38
4a.2	20.35±0.27	89.79
4b	18.31±0.30	89.80
A. flavus	25.00	83.19
C. graminicola	25.00	74.13
E. turcicum	25.00	74.08
F. graminearum	25.00	74.13
F. verticillioides	25.00	73.74
Cmn[c]	25.00	NA
M. phaseolina	25.00	74.30
D. pinnea	25.00	73.99
S.maydis	25.00	74.00
Phoma spp.	19.95±0.26	83.28
Water	25.00	73.76

Figure 4. Melting curve analysis for *Stenocarpella macrospora*using real-time polymerase chain reaction (PCR) assay. **A,** Amplification curves of *S. macrospora*isolates and controls obtained with primers RT.Smac.F/ RT.Smac.R. **B,** Melting curve analysis of the same samples shows the presence of the specific PCR product. **C,** Comparison chart with cycle threshold (C_t) values and melting temperatures. Superscript a: DNA extracted for other fungi isolated that are not the target of this assay and water are identified as negative controls. Superscript b: C_t values ± standard deviation. Superscript c: Cmn = *Clavibacter michiganensis* subsp. *nebraskensis*.

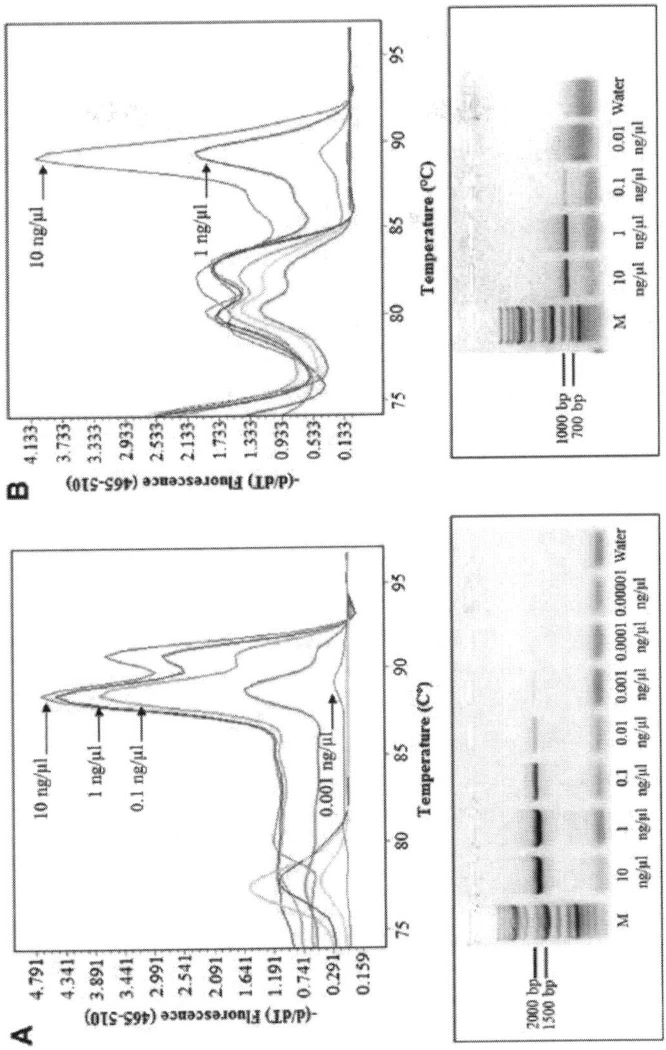

Figure 5. A, Dilution series for *Stenocarpella maydis* real-time (top image) and conventional (bottom/gel image) polymerase chain reaction (PCR) assays. Ten-fold dilutions were performed in each assay, with an initial concentration of 10 ng/µl to a final concentration of 0.00001 ng/µl. Gel image includes lane M: molecular size marker (Life Technologies, Grand Island, NY) and nuclease-free sterile water as control. **B,** *S. macrospora* real-time (top image) and conventional (bottom/gel image) PCR assays. Ten-fold dilutions were performed in each assay, with an initial concentration of 10 ng/µl to a final concentration of 0.01 ng/µl. Gel image includes lane M: molecular size marker (Life Technologies) and nuclease-free sterile water as control.

Real-time assays developed and validated here demonstrate that these techniques can be used to assess *S. macrospora* and *S. maydis* presence in both pure culture and in infected plant tissue. In this assay, two corn pathogens, *A. flavus* and *E. turcicum*, had slightly higher C_t values (26.08 and 27.21, respectively) than the range of *S. maydis* isolates (23.18 to 26.16). In the *S. macrospora* assay, *Phoma* spp. had a C_t value of 19.95, which is in the C_t range of *S. macrospora* isolates (18.31 to 21.69). However, in both assays, the identity of each respective organism was easily differentiated from nontarget pathogens by examining the melting profile of the amplified products. When SYBR Green is used for detection, it is difficult to determine whether the C_t value obtained is due to nontarget amplicons or to nonspecific products. Because of this, identification of isolates using these protocols should be based only on the analysis of the melting curves. Melting curves and peak locations clarify target species as positive samples and nontarget species as negative samples. For instance, melting curve peaks for *A. flavus* and *E. turcicum* are 64.25 and 77.73°C, respectively, which are lower than the range of *S. maydis* isolates (88.54 to 89.94°C). *Phoma* spp. had a melting curve peak of 83.28°C, compared with the range of 89.39 to 89.80°C for *S. macrospora* isolates. Schroeder et al. (43) described the advantage that real-time PCR offers by observing the melting curves of each product. Melting curves are based on the GC amount, length, and sequence that each sample contains and, for that reason, melting curve peaks for the target *Stenocarpella* spp. may vary slightly when implemented in other laboratories (39).

SYBR Green was selected for use in the real-time assays due to the relative simplicity and low cost of the product (4,5,34). Buh Gašparič et al. (4) compared nine different real-time chemistries and concluded that SYBR Green is suitable for use in both qualitative and quantitative detection assays. However, SYBR Green binds with all double-stranded DNA, including nontarget amplicons as well as nonspecific products. This can result in florescence in samples that are free of the target pathogen (39). This could explain why florescence was observed in certain real-time PCR assays of nontarget organisms for which no amplicon was detected in conventional PCR assays. This potential disadvantage is remedied by examining the melting profile at the end of the real-time assay (39,43). Other assays using real-time PCR employing the fluorescent dye SYBR Green for the detection and diagnosis of plant pathogens have shown that the analysis of the melting profile allowed differentiation of the target with the nontarget species. Tessitori et al. (49) was able to confirm and differentiate two different citrus viroids, *Citrus exocortis viroid* and Citrus viroid-IIb, by the use of melting temperatures and their melting peaks. Torres et al. (50) also described the advantage of using the melting curve analysis to differentiate three different apple proliferation phytoplasmas (*'Candidatus* Phytoplasma pyri', *'Ca.* P. pronorum', and *'Ca.* P. mali') by their melting peaks. Therefore, we concluded that the use of SYBR Green and the analysis of the melting profile using real-time PCR is a reliable tool for the diagnosis of *Stenocarpella* spp.

S. maydis and *S. macrospora* were identified directly from corn kernels with both conventional and real-time PCR assays. Amplified bands were reliable enough to confirm the presence of both *Stenocarpella* spp. Positive confirmation in infected corn seed using the real-time assay was based on the melting peak observed in the melting curve analysis. Samples were concluded to be positive for *S. maydis* and *S. macrospora* if the melting peak was located at 88 and 89°C, respectively. The use of both assays on DNA samples extracted from infected corn seed simplifies the identification of this pathogen and increases the utility of these assays in seed-monitoring programs. However, weak bands were observed in the conventional PCR assay when DNA was extracted directly from corn leaves infected with *S. macrospora*. Because these amplified bands were not as strong as those observed when *S. macrospora* DNA was amplified directly from corn kernels or from fungal cultures, we cannot reliably confirm *S. macrospora* directly from leaf tissue with the conventional PCR assay. However, both conventional and real-time PCR assays were able to reliably detect *S. macrospora* fungal presence at a detection limit of 0.1 and 1 ng/µl, respectively, from fungal culture, and 10 and 1 ng/µl, respectively, from infected corn seed. Detection of fungal presence in corn seed using DNA extracted from cultures or directly from infected tissue would be the primary use of these assays in regulatory agencies testing for seed quality, and incubation of infected tissue to increase the amount of DNA extracted from the pathogen may improve sensitivity in the assay (2). Further testing is needed to determine whether amplification of *S. macrospora* from infected leaves can be achieved with a conventional PCR assay.

Infected corn seed has been identified as an important source of inoculum for both*Stenocarpella* spp. (7,26,27), and can be regulated due to the potential for importing mycotoxin-affected grain into an area (11,16,28,32,38). Barrocas et al. (2) described the necessity of having a specific and accurate method for the identification of both*Stenocarpella* spp. for the Brazilian seed certification program, where *S. maydis* is listed as an RNQP and the Ministry of Agriculture has proposed that imported seed lots of certified seed may not contain more than 2% *S. maydis* (Administrative Regulation Number 47 of 19 February 2009). Inaccurate or unreliable diagnostic methods can negatively impact seed commercialization. The assays developed in this study can be useful tools in areas where *S. maydis* has been categorized as an RNQP.

ACKNOWLEDGMENTS

We thank C. P. Woloshuk for providing access to the *S. maydis* genome and for providing comments on the manuscript; G. Bergstrom, C. Bradley, T. Hughes, L. Sweets, and H. Young-Kelly for providing isolates; M. C. Aime for helpful discussion during the development of this study; N. Anderson for laboratory assistance and comments on the manuscript; and R. Koch and Z. Sexton for valuable comments and suggestions on the manuscript. Financial support was provided by the United States Department of Agriculture NIFA AFRI grant number 2010-85117-20607.

REFERENCES

1. Adipala, E., Lipps, P. E., and Madden, L. V. 1993. Reaction of maize cultivars from Uganda toExserohilum turcicum. Phytopathology 83:217-223.

2. Barrocas, E. N., Machado, J. C., Almeida, M. F., Botelho, L. S., and Von Pinho, É. V. R.2012. Sensibility of the PCR technique in the detection of Stenocarpella sp. associated with maize seeds. Rev. Bras. Sementes 34:218-224.

3. Bradley, C. A., Pedersen, D. K., Zhang, G. R., and Pataky, N. R. 2010. Occurrences of diplodia leaf streak caused by Stenocarpella macrospora on corn (Zea mays) in Illinois. Plant Dis. 94:1262.

4. Buh Gašparič, M., Tengs, T., La Paz, J., Holst-Jensen, A., Pla, M., Esteve, T., Žel, J., andGruden, K. 2010. Comparison of nine different real-time PCR chemistries for qualitative and quantitative applications in GMO detection. Anal. Bioanal. Chem. 396:2023-2029.

5. Cao, H., and Shockey, J. M. 2012. Comparison of TaqMan and SYBR Green qPCR methods for quantitative gene expression in tung tree tissues. J. Agric. Food Chem. 60:12296-12303.

6. Casa, R. T., Reis, E. M., and Zambolim, L. 2006. Doenças do Milho Causadas por Fungos do Gênero Stenocarpella. Fitopatol. Bras. 31:427-439.

7. Casa, R. T., Reis, E. M., and Zambolim, L. 1988. Fungos associados a semente de milho produzida nas Regioes Sul e sudesde do Brasil. Fitopatol. Bras. 23:370-373.

8. Casa, R. T., and Zambolim, L. 1997. Diplodia maydis e Diplodia macrospora associados à semente de milho. Dissertação de Mestrado, Universidade Federal de Viçosa, Viçosa, Brazil.

9. Crous, P. W., Slippers, B., Wingfield, M. J., Rheeder, J., Marasas, W. F. O., Philips, A. J. L., Alves, A., Burgess, T., Barber, P., and Groenewald, J. Z. 2006. Phylogenetic lineages in the Botryosphaeriaceae. Stud. Mycol. 55:235-253.

10. Cutler, H. G., Crumley, F. G., Cox, R. H., Cole, R. J., Dorner, J. W., Latterell, F. M., andRossi, A. E. 1980. Diplodiol: A new toxin from Diplodia macrospora. J. Agric. Food Chem. 28:135-138.

11. Darvall, P. M. 1964. Mouldy corn cobs, a danger to cows. Queensl. Agric. J. 90:692-693.

12. Earle, F. S. 1897. New species of fungi imperfecti from Alabama. Bull. Torrey Bot. Club 24:28-32.

13. Eddins, A. H. 1930. Dry rot of corn caused by Diplodia macrospora Earle. Phytopathology 20:439-447.

14. Gardes, M., and Bruns, T. D. 1993. ITS primers with enhanced specificity for basidiomycetes - application to the identification of mycorrhizae and rusts. Mol. Ecol. 2:113-118.

15. Johann, H. 1935. Diplodia macrospora on corn in Brazil. Plant Dis. Rep. 19:9-10.

16. Kellerman, T. S., Prozesky, L., Schultz, R. A., Rabie, C. J., Van Ark, H., Maartens, B. P., and Lübben, A. 1991. Perinatal mortality in lambs of ewes exposed to cultures of Diplodia maydis (= Stenocarpella maydis) during gestation. Onderstepoort J. Vet. Res. 58:297-308.

17. Kellerman, T. S., Rabie, C. J., Van der Westhuizen, G. C., Kriek, N. P., and Prozesky, L.1985. Induction of diplodiosis, a neuromycotoxicosis, in domestic ruminants with cultures of indigenous and exotic isolates of Diplodia maydis. Onderstepoort J. Vet. Res. 52:35-42.

18. Kriek, N. P. J., and Marasas, W. F. O. 1979. Toxicity of Diplodia macrospora to laboratory animals. Food Cosmet. Toxicol. 17:233-236.

19. Latterell, F. M., and Rossi, A. E. 1983. Stenocarpella macrospora (=Diplodia macrospora) and S. maydis (=D. maydis) compared as pathogens of corn. Plant Dis. 67:725-729.

20. Latterell, F. M., Rossi, A. E., and Moreno, R. 1976. Diplodia macrospora: A potentially serious pathogen of corn in the U.S.? (Abstr.) Proc. Am. Phytopathol. Soc. 3:28.

21. Lievens, B., and Thomma, B. P. H. J. 2005. Recent developments in pathogen detection arrays: Implications for fungal plant pathogens and use in practice. Phytopathology 95:1374-1380.

22. Lipps, P. E., and Mills, D. R. 2002. Diplodia ear rot of corn. The Ohio State University Extension Fact Sheet AC-0046-01. Online publication.

23. Marasas, W. 1977. The genus Diplodia. Pages 119-128 in: Mycotoxic Fungi, Mycotoxins, Mycotoxicoses. An Encyclopedic Handbook, Vol. 1. T. D. Wyllie and L. G. Morehouse, eds.Marcel Decker, Inc., NY.

24. Marasas, W. F. O., and Van der Westhuizen, G. C. A. 1979. Diplodia macrospora: The cause of a leaf blight and cob rot of maize (Zea mays) in South Africa. Phylophylactica 11:61-64.

25. Mario, J. L., and Reis, E. M. 2001. Método simples para diferenciar Diplodia macrospora de D. maydis em testes de patologia de sementes de milho. Fitopatol. Bras. 26:670-672.

26. McGee, D. C. 1988. Maize Diseases: A Reference Source for Seed Technologists. American Phytopathological Society, St. Paul, MN.

27. McNew, G. L. 1937. Crown infection of corn by Diplodia zeae. Iowa Agric. Exp. Stn. Res. Bull. 216. Ames.

28. Mitchell, D. T. 1919. Poisoning of cattle by Diplodia-infected maize. S. Afr. J. Sci. 16:446-452.

29. Morant, M. A. 1988. Stenocarpella (Diplodia) maydis: Influence of nitrogen and inoculation on biochemical components of maize stalk. Ph.D. diss. Botany and Plant Pathology Department, Purdue University, West Lafayette, IN.

30. Morant, M. A., Warren, H. L., and Vonqualen, S. K. 1993. A synthetic medium for mass-production of pycnidiospores of Stenocarpella maydis. Plant Dis. 77:424-426.

31. Mueller, D., and Wise, K. A. 2012. Corn disease loss estimates from the United States and Ontario, Canada—2012. Purdue Ext. Publ. BP-96-12-W. Online publication.https://www.extension.purdue.edu/extmedia/BP/BP-96-12-W.pdf

32. Odriozola, E., Odeon, A., Canton, G., Clemente, G., and Escande, A. 2005. Diplodia maydis: A cause of death of cattle in Argentina. N. Z. Vet. J. 53:160-161.

33. Pascual, C. B., and Relevante, C. A. 2003. Cultural and molecular characterization ofStenocarpella macrospora (Earle) Sutton causing leaf blight, ear rot and stalk rot in maize. J. Trop. Plant Pathol. 39:36-42.

34. Ponchel, F., Toomes, C., Bransfield, K., Leong, F. T., Douglas, S. H., Field, S. L., Bell, S. M., Combaret, V., Puisieux, A., and Mighell, A. J. 2003. Real-time PCR based on SYBR-Green I fluorescence: An alternative to the TaqMan assay for a relative quantification of gene rearrangements, gene amplifications and micro gene deletions. BMC Biotechnol. 3:18.

35. Rao, S. K., and Achar, P. N. 2001. Screening and in vitro production of diplodiatoxin from the isolates of Stenocarpella maydis and its toxigenic effect on bacterial strains. Indian J. Exp. Biol. 39:1243-1248.

36. Reis, E. M., Casa, R. T., and Bresolin, A. C. R. 2004. Manual de diagnose e controle de doenças do milho, 2nd ed. Graphel Gráfica e Editora Lages, Santa Catarina, Brazil.

37. Riet-Correa, F., Mendez, M. C., and Schild, A. L. 1984. Page 35 in: Laboratório Regional de Diagnóstico. Doenças diagnosticadas no ano 1983. Editora da Universidade, Pelotas, Brazil.

38. Riet-Correa, F., Schild, A. L., and Fernandes, C. G. 1998. Enfermidades do sistema nervoso dos ruminantes no sul do Rio Grande do Sul. Cienc. Rural 28:341-348.

39. Ririe, K. M., Rasmussen, R. P., and Wittwer, C. T. 1997. Product differentiation by analysis of DNA melting curves during the polymerase chain reaction. Anal. Biochem. 245:154-160.

40. Romero Luna, M. P.2012. Managing Diplodia ear rot in corn: Short and long-term solutions. M.S. thesis, Botany and Plant Pathology Department, Purdue University, West Lafayette, IN.

41. Rossouw, J. D., Pretorius, Z. A., Silva, H. D., and Lamkey, K. R. 2009. Breeding for resistance to Stenocarpella ear rot in maize. Plant Breed. Rev. 31:223-246.

42. Saghai-Maroof, M. A., Soliman, K. M., Jorgensen, R. A., and Allard, R. A. 1984. Ribosomal DNA spacer length in barley: Mendelian inheritance, chromosomal location, and population dynamics. Proc. Natl. Acad. Sci. USA 81:8014-8018.

43. Schroeder, K. L., Okubara, P. A., Tambong, J. T., Lévesque, C. A., and Paulitz, T. C. 2006.Identification and quantification of pathogenic Pythium spp. from soils in Eastern Washington using real-time polymerase chain reaction. Phytopathology 96:637-647.

44. Siqueira, C. S., Barrocas, E. N., Machado, J. C., Silva, U. A., and Dias, I. E. 2014. Effects ofStenocarpella maydis in seeds and in the initial development of corn. J. Seed Sci. 36:79-86.

45. Snyman, L. D., Kellerman, T. S., Vleggaar, R., Flett, B. C., Basson, K. M., and Schultz, R. A. 2011. Diplonine, a neurotoxin isolated from cultures of the fungus Stenocarpella maydis (Berk.) Sacc. that induces diplodiosis. J. Agric. Food Chem. 59:9039-9044.

46. Steyn, P. S., Wessels, P. L., Holzapfel, C. W., Potgieter, D. J. J., and Louw, W. K. A. 1972.The isolation and structure of a toxic metabolite from Diplodia maydis (Berk.) SACC. Tetrahedron 28:4775-4785.

47. Sutton, B. C., and Waterston, J. M. 1966. Diplodia macrospora. CMI Descriptions of pathogenic fungi and bacteria. Set 9. No. 83. CAB International, Wallingford, UK.

48. Sutton, B. C., and Waterston, J. M. 1966. Diplodia maydis. CMI Descriptions of pathogenic fungi and bacteria. Set 9. No. 84. CAB International, Wallingford, UK.

49. Tessitori, M., Rizza, S., Reina, A., and Catara, V. 2005. Real-time RT-PCR based on SYBR-Green I for the detection of citrus exocortis and citrus cachexia diseases. Pages 456-459 in: Proc. 16th Conf. Int. Organ. Citrus Virol. M. E. Hilf, N. Duran-Vila, and M. A. Rocha-Peña, eds. IOCV, Riverside, CA.

50. Torres, E., Bertolini, E., Cambra, M., Montón, C., and Martín, M. P. 2005. Real-time PCR for simultaneous and quantitative detection of quarantine phytoplasmas from apple proliferation (16SrX) group. Mol. Cell. Probes 19:334-340.

51. Vincelli, P. 1997. Ear rot of corn caused by Stenocarpella maydis (=Diplodia maydis). Univ. Kentucky Coop. Ext. Serv. PPA-43.

52. Vincelli, P. 2011. Diplodia leaf streak of corn. Pages 2-3 in: Kentucky Pest News No. 1277. Univ. Kentucky Coop. Ext. Serv.

53. White, D. G. 1999. Page 44 in: Compendium of Corn Diseases, 3rd ed. American Phytopathological Society, St. Paul, MN.

54. White, T., Bruns, T., Lee, S., and Taylor, J. 1990. Amplification and direct sequencing of fungal ribosomal RNA genes for phylogenetics. Pages 315-322 in: PCR Protocols: A Guide to Methods and Applications. M. Innis, D. Gelfand, J. Shinsky, and T. White, eds. Academic Press, New York.

55. Wicklow, D. T., Rogers, K. D., Dowd, P. F., and Gloer, J. B. 2011. Bioactive metabolites fromStenocarpella maydis, a stalk and ear rot pathogen of maize. Fungal Biol. 115:133-142.

56. Wise, K. A. 2011. Northern corn leaf blight. Diseases of corn. Purdue Univ. Purdue Ext. BP-84-W.

57. Xia, Z., and Achar, P. N. 2001. Random amplified polymorphic DNA and polymerase chain reaction markers for the differentiation and detection of Stenocarpella maydis in maize seeds. J. Phytopathol. 149:35-44.

Index